Successful Engineering

A Guide to Achieving Your Career Goals

Lawrence J. Kamm

McGraw-Hill Book Company

New York St. Louis San Francisco Auckland Bogotá
Caracas Colorado Springs Hamburg Lisbon
London Madrid Mexico Milan Montreal
New Delhi Oklahoma City Panama Paris
San Juan São Paulo Singapore
Sydney Tokyo Toronto

Library of Congress Cataloging-in-Publication Data

Kamm, Lawrence J.
    Successful engineering: a guide to achieving your career goals
    Lawrence J. Kamm.
        p. cm.
    Includes index.
    ISBN 0-07-033267-3
    1. Engineering--Vocational guidance.   I. Title.
TA157.K25 1989
620'.0023--dc19

1234567890   DOC/DOC   8954321098

ISBN 0-07-033267-3

*The editors for this book were Robert Hauserman and Beatrice E.
Eckes, the designer was Naomi Auerbach, and the production
supervisor was Richard Ausburn. It was set in Century Schoolbook. It
was composed by the McGraw-Hill Book Company Professional &
Reference Division composition unit.*

Printed and bound by R. R. Donnelley & Sons Company.

*For more information about other McGraw-Hill materials,
call 1-800-2-MCGRAW in the United States. In other countries,
call your nearest McGraw-Hill office.*

To Jacob Rabinow, my boss, teacher, close friend, and severest critic

Contents

Preface

This book is for professional engineers in all fields. Senior engineers who took the course on which this book is based have told me that it filled in and organized their own experience and have recommended it for other graduate engineers.

Its purpose is to teach you how to be successful in the real world of engineering after you have learned mathematics, science, and computational engineering in engineering school. If I had read this book when I graduated instead of learning it a little at a time over 47 years, I would have been more successful, richer, happier, and all much sooner.

For you "to be successful," I mean you:

- Produce the best designs of which you are capable
- Cause your proposed designs to be accepted and used
- Become as well rewarded as you can in money, position, and security
- Enjoy your professional life

The book tells what I have learned in 47 years of engineering and entrepreneuring. Most of its subject matter, examples, and anecdotes are taken from my own career. Other examples are taken from the career of Jacob Rabinow of the National Bureau of Standards, who is one of the great inventors of the country with over 200 patents. I worked for him during his foray into private enterprise and learned, and continue to learn, more from him than from anyone else I have ever met. He has been kind enough to read the manuscript of this book and make corrections.

The book emphasizes conceptual design rather than analytical design because my own aptitudes and history have had this emphasis and because I have nothing new to contribute to analytical design. However, most of the book deals with problems which are equally the concern of the purely analytical engineer and the engineer who produces conceptual designs, properly untroubled by how much they are "inventions."

Portions of the book give you information and ideas; other portions recommend actions I think you should take; but some portions recite problems for which I know no clear answers (e.g., ethics). These are presented for you to think about so you will be prepared to face them when they appear, with fewer shocks than I had.

There are theoreticians who believe that good design can be performed by a computer having a large database and an artificial-intelligence program. This practitioner is waiting for the program to be written which has the attributes of insight, judgment, persuasion, will, prediction, and ingenuity which produce successful designs by humans and which can "process" the kinds of human-behavior "data" contained in these chapters. One purpose of this book is to encourage you to further develop these attributes to increase your own degree of success.

In the illustrative anecdotes, names have been concealed to protect me from the guilty.

The book is written with the words "you" and "I" and is in as simple English as I can write. Please do not confuse pedantic complexity with philosophical profundity.

Design engineers work in manufacturing companies, consulting companies, government laboratories, and university laboratories. I use the word "company" as a generic for all these organizations.

I think you will come out of this book working more successfully than when you went in. Good luck!

LARRY KAMM
San Diego, California
1988

Introduction

When you design a product you must do a great deal more than solve the technical problems taught in engineering school. You must deal with a great range of problems from broad concepts to minute details together with problems which are not technical engineering at all.

You must suit the design to the peculiarities of your own company (12) and to the peculiarities of your customers (13). (Parentheses give the chapter numbers which discuss the subjects in detail.) You must make your design superior to the present and future designs of your competitors (14). You must learn all applicable general specifications and help to develop the specific specification for your product (19). You must research technical knowledge you do not yet have (9, 11) and call in consultants for aid when you decide that it would cost too much time for you to acquire certain knowledge and skills, including artistic-design skills (10). You must choose design options suitable for the quantities in which your product will be made (21). You must at all phases of the design consider the costs of the product and of the design effort (20). You must design the product to be appropriate for the maintenance and reliability ground rules which will apply (22). You must decide when to transfer your efforts from paper design to models and experiments and back to paper (23). You must consider all design objectives, not just those written in the specifications (25). You must design quantity products to be suitable for mechanized and automatic manufacturing (27). You must design your product to be suitable to the humans who will deal with it (29). And you must engineer the packaging so that it can be shipped to your customers (32).

Your design may be of a substantially standard structure or device which is better than its predecessors because of better materials and components and better mathematical analysis, or it may be better because it incorporates new concepts, ideas, or inventions, or both (1, 2).

You must produce documentation to tell your factory how to make the product, your customers how to use and to maintain the product, and your lawyers how to patent it and defend it from patent and product liability lawsuits.

You may have to help design some of the manufacturing and test equipment to produce the product and special maintenance tools and equipment for your customer to use.

After the product has been shipped, you may have to help your customer with problems associated with it and, unless you have done a phenomenally good job, you will have to make design changes in the product after the first units have been delivered.

As the product is produced, you will be called upon to make a seemingly endless series of design changes, usually in details, to correct design errors and to accommodate manufacturing, customer, and vendor problems. You must face the exasperating question, "These parts were made out of spec, but it will be expensive and time-consuming to reject them, so can we use them anyway?"

Finally you must accomplish all of this in an environment of people who supervise, cooperate with, or help you (3, 4), you hope, and in which you are furthering your career (5, 6).

The chapters in this book discuss these aspects of the real world of design engineering and will help you to cope with them better and sooner.

Successful Engineering

Innovation and Conceptual Design

Quantitative Design and Qualitative Design

Engineering Education

Engineering education deals primarily with calculating the *quantitative* performance of known engineering objects and with designing conventional variations in those objects. (By "engineering objects" I mean machines, circuits, dams, or any other things designed by engineers.)

In science courses we learn to compute the behavior of nature in a series of established experiments and phenomena. In engineering courses we learn to compute currents and voltages in amplifiers and motors, forces and power in linkages and engines, reaction rates in chemical systems, earth transfer in road building, and the like. This is good educational policy; learning applied mathematics is difficult for most of us and is best done in the disciplined operation of a school rather than left to be picked up on the job.

The ability to compute separates the engineer from the technician. An education in engineering mathematics generates an insight (i.e., an intuitive understanding) into the behavior of physical things which cannot be attained in any other way and which is essential for inventive thinking, that is, the generation of new *qualitative* ideas which will work. If you have really learned what calculus means, you have a gut feel for the behavior of billiard balls, automobiles, electric currents, servomechanisms, space vehicles, and all the other objects of engineering which no amount of practical experience alone can provide. On the other hand, *quantitative* design can be used only on an engineering object which originated as a *qualitative* idea.

To be successful, engineers need a large body of knowledge and skills that is not subject to mathematical computation; this body of knowledge and skills is the subject of this book. It deals with a variety of nonquantitative design information, ideas, and techniques which will help you to devise new and advanced engineering objects in your field.

The book also presents ideas and suggestions for dealing with other people and organizations with which you must work so that your technical engineering efforts produce designs which get developed, built, and sold. If you succeed in causing your designs to be so used, you will be personally better off for having done so.

Real-World Design Process

This process iterates around three elements:

1. *Qualitative design.* The generation of ideas, structures, concepts, combinations, configurations, and patterns. The results are expressed in sketches, layouts, schematics, and diagrams.

2. *Quantitative design.* The computation of the magnitude of the elements in a qualitative design. The results are expressed in numbers, usually with physical units (e.g., length, voltage, temperature).

3. *Experimental design.* The use of physical models and tests to compensate for both qualitative and quantitative uncertainty.

Designers start with some qualitative ideas, then calculate approximate quantitative magnitude, then revise the qualitative ideas as a result of deeper understanding produced by their calculations, then iterate between the two processes until they are satisfied with both ideas and magnitudes. The calculations may be quite approximate in the first few cycles and become as exact as desired (and possible) at the end. Physical experiments along the way compensate for both qualitative and quantitative uncertainties and sometimes may replace a more costly quantitative study with a less costly physical test.

Quantitative understanding, not just passing examinations, of science and mathematics gives you insight into the behavior of nature (including your own qualitative ideas) and helps you to generate workable qualitative designs.

Qualitative design is sometimes called synthesis; quantitative design is sometimes called analysis.

Uses and Limits of the Computer in Design

The computer is the greatest tool ever developed for computational engineering and, perhaps, for design drafting (but see the discussion on do-it-yourself CAD in Chap. 23). Efforts are under way by academic theoreticians, under the names of expert systems and artificial intelligence, to computerize qualitative design (see Chap. 28). This book deals with your use of insight, judgment, persuasion, will, prediction of human behavior, and ingenuity and with a number of principles of design to produce qualitative designs and get them used. Perhaps these capabilities and activities will be reduced to computer programs someday, but this design practitioner is not holding his breath.

Role of Human Judgment

The role of human judgment in design appears in predicting the acceptability of the design to other people and in predicting the performance of the design long before it is reduced to mathematical computation. For any problem there are many bad designs possible and only a few good ones. Human judgment is the first filter in selecting the good ones.

Challenger Fiasco

A glaring example of a lack of qualitative understanding of an engineering component in the presence of enormous quantitative capability was the *Challenger* disaster in 1986.

The *Challenger* blew up because the hot gases in a solid-fuel booster rocket burned through an O-ring seal. The plume of 6000°F gas then burned through the wall of the orbiter liquid-hydrogen tank. Escaping hydrogen reduced tank pressure, which permitted the adjacent oxygen tank wall to collapse. Then 160 tons of liquid oxygen and liquid hydrogen mixed, adjacent to the flames from the rocket engines, with the consequent white flash on our TV screens and national disgrace.

Figure 1.1 shows how an O ring works. An O ring is a skinny rubber

Figure 1.1   How an O ring works. (*a*) As installed. (*b*) Under pressure. (*From* The Parker O-Ring Handbook, *courtesy of Parker Seal Group.*)

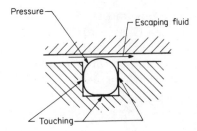

Figure 1.2 Leakage past a narrow O-ring groove.

doughnut squeezed into a groove between parts which are to be sealed. Pressure from the sealed fluid pushes the O ring ahead of it into the gap between the body parts so that the O ring obstructs passage of the gas. This is called a self-energizing seal. The gas must exert pressure on the entire left side of the O ring or, instead of pushing it forward and upward to block the escape route, the gas will push it down, out of the way of the escape route, and the gas will escape. (See Fig. 1.2.) Therefore the O-ring groove must be wider than the compressed O ring; otherwise, the O ring will touch all four sides of its enclosure and will not seal.

Standard dimensions for O rings and their grooves have been established for at least 30 years, are common handbook data, and are available in free handbooks from O-ring manufacturers.

With the aid of finite element analysis the President's Commission to investigate the disaster[1] demonstrated that, with the groove dimensions designed, if the metal-to-metal gap got as small as 0.004 in, there would be four-point contact, the pressure would not apply to the entire side, and self-energizing would not occur.

The two parts sealed were rocket body segments 12 ft in diameter and ½ in thick. It would be obvious to any competent junior engineer that, because of imperfect roundness, there must have been points where the gap was 0.000 in, not 0.004 in. Furthermore, there is no recognition in the report that this was not the first O-ring seal ever built and that the design should have been an elementary-handbook exercise.

Figure 1.3 shows a nominal O-ring diameter of 0.28 in which is approximately the same as the standard diameter of 0.275 in. It shows a groove width of 0.310 in, but the handbook width is 0.375 to 0.380 in, or more than 0.065 in wider!

Thus a lack of qualitative understanding of what they were doing caused the seal designers to kill seven people and wreck a multibillion-dollar program.

Furthermore, the lack of elementary qualitative knowledge of common O-ring practice by the investigators on the commission made

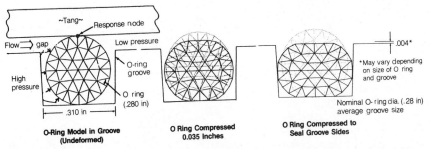

Figure 1.3 *Challenger* O-ring analysis in the *Report to the President by the United States Presidential Commission on the Space Shuttle Challenger Accident. (Government Printing Office.)* Drawings show how progressive reductions of the gap between tang and clevis can inhibit and eventually block the motor cavity's high-pressure flow from getting behind the O ring.

them go through the exercise of reinventing O-ring theory with a computer finite element analysis and never recognizing that the *Challenger* designers had violated an old-established principle of design.

I will have more to say about this low point in the history of American technology in Chap. 15, "Simplicity."

REFERENCES

1. *Report to the President by the United States Presidential Commission on the Space Shuttle Challenger Accident* (the Rogers Report), Superintendent of Documents, Washington, June 6, 1986.

Textbooks on quantitative design

2. Dieter, George E.: *Engineering Design,* 1st ed., McGraw-Hill, New York, 1983.
3. Pahl, G., and W. Beitz: *Engineering Design,* 1st ed., ed. by K. Wallace, Springer-Verlag, New York, 1984.

Inventing

Design engineers conceive new ideas, some of which meet the criteria of patentable invention and some of which do not. A patentable invention may, very loosely, be described as a new idea which would not be developed by an expert in the routine practice of a profession. More rigorous definitions appear in the references. A nonpatentable invention is a new idea which does not meet both criteria (i.e., that it be new and not routine). From the standpoint of the design engineer there is no difference except the business and prestige value of patents. In this book I have used the word "invention" to mean any good design idea you think up. Inventions extend over a wide spectrum of ingenuity from ho-hum to brilliant.

Patents

References 3, 8, 16, 25, and 28 give more rigorous and detailed definitions of patentability and introductory descriptions of patent law and practice.

The usual procedure in obtaining a patent is to hire a patent attorney or agent to order a search and then to prepare the application and prosecute it in the U.S. Patent and Trademark Office until the patent is issued. The costs run from $2000 to $20,000. Reference 14 is a list of registered patent attorneys and agents.

If you are a frequent inventor, it is worthwhile to take a university extension course in patent law (approximately two terms at one night per week) and then prepare and prosecute your own applications. At the very least, you will understand the system better and communicate with your patent attorney better. You will also be qualified to take the Patent Office examination and be registered as a patent agent.

The same examination is given to both lawyers and engineers. At one time both earned the title "Patent Attorney," but the lawyers then invented and sold the semantic ploy that engineers be called "Patent Agents" and only lawyers be called "Patent Attorneys." To a layperson it would appear that a patent attorney outranks a patent agent, so the lawyers won a competitive advantage. It makes no difference whatever to the Patent and Trademark Office, but only lawyers can argue infringement cases which go to court.

I followed this route and filed a number of patent applications by myself. For the few cases in which it developed that the invention was of commercial value I then hired a professional to amend the application so that a stronger patent would result. (The professional I hired was my patent law professor.) In cases in which I decided that there would be no commercial value in the patent I just abandoned the application.

For your application you will need patent drawings, which must be made in accordance with Patent Office specifications.[10] You can hire a patent drafter directly, or you can learn to do the work yourself if you have the skill.

Purpose of the patent system

Why do we have a patent system, and what does it do? The purpose of the patent system, in the public interest, is to encourage inventors both to invent and to disclose their inventions to the public to provide a basis for further inventing by others. To accomplish this the government offers a deal to the inventor. If the inventor discloses enough about an invention so that others can reproduce it, then the government rewards him or her with a limited monopoly of the use of the invention.

What is a patent?

The patent document discloses the invention in the "specification" and "drawings" and defines the scope of the monopoly in the "claims." The "monopoly" is actually the right to exclude others from using the invention rather than a right to use it yourself. For example, if you already have a patent on the automobile and I then patent an improved wheel, I can prevent you from making automobiles with my improved wheel but you can prevent me from making my improved wheel as part of an automobile. In practice, we then get together and cross-license.

Patents as an educational resource

Despite this arrangement the patent system is neglected as a source of education by most American design engineers. After World War II the United States sent a group of scientists to Europe in Operation Paper Clip to round up German scientists and engineers, particularly those responsible for the V-1 and V-2 guided-missile programs. (The Russians, of course, ran a matching program.) We netted Wernher von Braun and most of the engineers and scientists who had worked with him at Peenemünde. When von Braun was asked how he developed the V-2 rocket in such a short time compared with the little progress which had been made in the United States, he answered, with surprise, "Why, we studied the patents of your great Dr. Goddard, of course."

If you become a serious inventor in a particular art, you will at first be disappointed that most of what you invent turns up on patent search as old art. However, as you become more and more the master of that art, you will find that the anticipating patents are more and more recent until you start finding that some of your inventions are new and patentable.

You will also learn that most patentable inventions are based on a recent problem or opportunity or a new material or technology. Most of what could have been invented in the past *was* invented in the past. It is discouraging to discover how many other smart people there have been and still are. The continuing occurrence of "interferences" (duplicate inventions by different inventors at about the same time) in the Patent and Trademark Office demonstrates this. Inventing is also a race, so you should maintain your sense of urgency and competition.

This section has been only a most superficial introduction to the patent system. If you are interested in patents, I urge you to write for the 41-page booklet *General Information Concerning Patents* listed as Ref. 8 among the other Patent Office publications.

Kinds of Invention

Invention, varying from clever improvements of details to brilliant breakthroughs of technology, spans a tremendous spectrum of merit. A better screwhead cavity is an invention in this sense, and so are the laser, the jet engine, and the transistor.

Some people spend their professional lives refining existing products without ever inventing anything and yet produce great value in their work, some people invent new products of great value, and some people invent all their lives and do not produce anything of value at all.

We engineers think of inventing as being a uniquely engineering process, but inventing is done in almost all fields of human activity:

- Scientists invent hypotheses and instruments.
- Authors invent characters and plots.
- Artists invent forms.
- Composers invent melodies and rhythms.
- Business people invent deals. (If you become an entrepreneur, you will have the opportunity to do so, as I have done, and will realize the wealth of opportunities which exists.)
- Soldiers invent tactics.
- Parents invent games.
- Some criminals invent frauds, and others invent techniques, tools, and tactics for violent crimes.
- Business lawyers invent contract terms, and litigating lawyers invent arguments and strategies.
- Surgeons invent procedures and instruments. (They sometimes work in cooperation with engineers, as an example in this book illustrates.)
- Diplomats invent treaty terms.
- Legislators invent laws.
- Financiers invent financing deals.
- Salespeople invent persuasive arguments.
- Marketers invent strategies.

Inventiveness

The word "creativity" is widely used to describe the aptitude and productivity of inventors. It is quite a proper word, but it is used so much in ways I do not entirely admire that I have developed a distaste for it and do not often use it. Aptitude, talent, productivity, inventiveness, and ingenuity are some of the words which I prefer and which I use here.

I believe that the ability to invent can be encouraged and developed, but I also believe that it is based on inborn aptitude, which varies from person to person just as many other aptitudes do. (When I say "I believe," I mean that I have an opinion based on my own observations, reading, conversations, and thought, but not established with scientific proof. If you disagree with my "belief" or call it a prejudice, I will not quarrel very hard.)

To illustrate my belief, I will tell you that I am the same age as Isaac Stern, the great violinist. We were both born to middle-class American families, he in San Francisco and I in New York. At an early age we each had a violin tucked under his chin by a doting mother. A year later I discarded my scratch box in what I still consider one of the happiest days of my life, and he went on to become Isaac Stern, the great violinist. There is a difference in talent.

I have known or heard of people who have unusual talents in music, art, invention, and abstract thinking in law and philosophy but only average talent elsewhere. I have also known and heard of people with multiple talents. The greatest are called Renaissance men after Leonardo da Vinci and Michelangelo. (There are also people with a great deal to be modest about.)

Commercial testing companies administer psychological tests to measure a variety of aptitudes such as spatial visualization, mathematics, and abstract thinking as well as other personality characteristics. (I was once sent for a full day of such testing before I was accepted for an interview. I must have done all right, because I got the interview and then a job offer.) I believe (there's that word again) that most of us would be better off if we were similarly tested before starting college so we could choose a career path for which we were well fitted. Our scores could be kept private to prevent their prejudicing other people against us.

When I hired engineers, I looked for inventive talent because it was appropriate for the work my companies did. I used a test, originated by Jack Rabinow, in which the applicant is asked to propose as many different designs as possible to solve a simple mechanism problem.

The very best test performance ever was by John Toth, who had just been denied promotion by the personnel department of major aerospace company A because he did not have an engineering degree from an accredited university. John became the most valuable engineer I ever hired, in both qualitative and quantitative design.

The very worst test performance ever was by an engineer who had an M.S. in M.E. degree from a prestigious university, was a highly respected engineer in a most prestigious high-technology company, and was available only because of a division closedown. He was the only person I ever interviewed who could not think of a single design idea. I found that his work had been stress analysis of complex structures. I had found (but didn't hire) a 100 percent pure quantitative design engineer.

What are the attributes of an inventor? The inventor mentally combines, proportions, and operates in his or her head the elements of an invention. Therefore the ability to visualize mechanical structures and their motions or electrical or chemical circuits and structures and their invisible actions is essential. The fisher's expression "to think like a fish" suggests the inventor's ability to think like the invention. The same ability enables the engineering diagnostician (debugger) to visualize the internal performance of a real device which is not operating as desired.

The most dramatic and important debugging visualization I ever did was on the Convair heart-lung machine I had designed (Fig. 2.1). It was during our second human operation. Just before connection to the patient, when the machine was circulating blood through a closed loop of tubing, great gobs of air suddenly appeared in the blood. Nothing like this had ever happened during dozens of development tests. The patient was wide open and too sick for the surgeon to close him up and try again. The machine was sterile and could not be opened for examination and modification.

I visualized the elements and operation of the machine and saw that the effect could come from an unsymmetrical collapse of one of the pumping bladders such that a fold of the bladder would block the exit hole. I then realized that the bladders came from a new batch and therefore might have some slight differences from the ones we had been using, thus supporting the hypothesis. Further visualizing the machine, I invented the idea that the bladders could be operated with

Figure 2.1 Heart-lung machine. (*Courtesy of General Dynamics Corporation.*)

only partial collapse if the valve gear of the water engine which pow-ered the machine were tripped by hand at half stroke. The procedure worked. Eddie Leak, steady as a rock, and I sat on the floor for 20 min-utes, reached into the mechanism with long screwdrivers, and tripped the linkages at half stroke every cycle, knowing that if we made a mistake we would kill the man on the table. He lived.

The engineering fix was easy: adding a perforated tube to limit bladder displacement. But I will be satisfied not to play Thurber's Walter Mitty again.

Developing your talent

To what degree are we confined to our genetic talent, and to what de-gree can we develop whatever talent we have? Talent alone produces nothing without development. I know of no 4-year-old inventor, and even Mozart took piano lessons.

I know a few rules which will help you.

1. *The hardest is brute-force education: school, homework, examina-tions, and grades.* Mathematics and science, if you really understand them and don't just pass examinations, teach you how nature works. (Inventions are new combinations of parts of nature.) Engineering courses teach you the existing elements and combinations of your craft and how to use your mathematics for their quantitative design. I'll present more of this later in the section called "Your Knowledge Base."

2. *Practice inventing.* (Remember Mozart and his piano.) If I may be pompous for a moment, the human mind is the only instrument which gets sharper the more you use it.

3. *Permit unconventionality within the privacy of your own head.* If you fear or dislike having a silly idea, or a disrespectful idea, or a dumb idea, or an immoral idea, then you won't have one. You also won't have the great idea which appears among a thousand poor ones. An invention, in its nature, is unconventional. Remember, I said, "in privacy." What you announce out loud is the result of your internal filtering, and you adjust your filter to match the sympathy and under-standing of the individuals you talk to.

Mental process of inventing

What happens inside the head of an inventor during the process of in-venting? I doubt if anyone knows for sure, but Jack Rabinow and I be-lieve it is approximately this:

As soon as a problem is considered, there is a stream of free associ-ation of ideas, existing devices, components, and materials and other memories of many kinds, including things quite unrelated to engineering. They appear in combinations and with variations in scale, shape, and material which seem relevant to the problem. The selection process is not conscious, but the elements selected become conscious. I do not know the mechanism by which a very small fraction of one's enormous store of memories becomes selected except the psychologist's word "association."

There is parallel thinking as well as serial thinking; i.e., the brain juggles many balls in the air at the same time. This is the opposite of the disciplined, logical thinking we engineers have worked so hard to learn.

The result is a jumble of schemes, most of which are worthless. (Remember what I said, above, about permitting unconventionality in the ideas inside your head.) Judgment enters, as you go along, and discards the very bad ideas or components of ideas. Then engineering rationality enters with approximate calculations and other criticisms of the schemes in mind. The judgment and the rationality do not appear at any programmed times—you use them when your feelings tell you to use them.

I said, "as soon as a problem is considered." Do *not* freeze your thinking by precisely defining the problem as you were taught to do in school. Modifying the *problem* may be a valuable part of the invention.

If you are talented and lucky and hard-thinking, a valuable invention, quite logical and practical, will rise out of the wild confusion in which it was born.

Everyone knows that the mind is at work unconsciously while we sleep and while we do unrelated things. Does it invent unconsciously? Yes, it does. Most of the process described above can also take place without our even knowing that it is going on. I do not suggest that to invent you should watch TV and just jot down the brilliant conceptions which interrupt the football game. Your unconscious thinking requires the same intensity of effort as your conscious thinking.

Personality traits of inventors

What are the personality traits of inventors? In the first place, inventors are unconventional, or they would not think up unconventional (i.e., innovative) devices. Thus they have widely different personalities. What specific traits and habits contribute to inventive productivity?

Hard work, meaning hard thinking, causes more associations and variations to occur than does relaxation. The thinking may take the form of compulsive brooding. Active reading of technical literature and observation of the physical world around you may provide elements without passage through memory.

Thinking about the problem during boredom time—shopping with spouse, attending dull concerts, driving, exercising, etc.—helps the process. Energetic people do more hours of hard thinking than do sluggish people.

Stubbornness and perseverance contribute to success in innovation, both in the mental struggle to generate the ideas to solve a problem and in physical experimentation. Edison was a paragon of persistent hard work. The story of his developing the high-resistance lamp filament by experiment after experiment until he was successful needs no repetition here. He is quoted as saying, "Genius is 1 percent inspiration and 99 percent perspiration."

Some people have what the psychologists call an "obsessive-compulsive personality." They brood on a problem over and over again, and if they also have talent, they produce results which a more lighthearted person with equal talent will not produce.

Some minds think better in concrete terms: wheels and volts and shapes and pH and geometry. Some minds think better in abstract terms: entropy and energy, and legality, and pure mathematics. Some minds think better in details, others in systems. Invention occurs in all these regimes, but the more the mind can encompass both concrete and abstract thinking and both detail and system thinking, the more and better innovations that can be produced.

There are passive personalities, who only respond to direction from others, and aggressive personalities, the self-starters, who act on their own initiative. The aggressive ones are both blessings and curses to their managers, but they are the ones who invent.

Many famous inventors have been aggressive both in their conceptual work—the actual inventing—and in the promoting of their inventions into use and commercial success. Edison, Ford, Firestone, McCormick, Tesla, and Marconi are a few of the names in this category.

Motivation.   What motivates people to invent? Perhaps there is some spiritual force of sheer creativity. Perhaps. But there is no perhaps about the forces of egotism, greed, security, and loyalty.

Every inventive person I have ever heard of, in any field, is an egotist, whether an outspoken boaster like Picasso or a shy introvert like Van Gogh. Bringing forth inventions establishes and maintains self-esteem and the respect of others. Patents and awards and publicity are prestige merit badges. Whether or not an inventor has a deep-seated insecurity which requires this form of reassurance I leave to the inventor's psychoanalyst, but the egotism is universal.

"Greed" is a dirty word, but I will stick to it. I include desire for money and desire for position. Ambition is a more complimentary term. Most inventors of engineered devices, business deals, and works of literature and art hope to make a lot more money from selling their creations than they can hope to make by earning a salary. The standard price for every amateur's invention is $1 million. There is some practical advice on the subject below under the heading "How to Benefit from Your Inventions."

Security, or at least the feeling of security, is helped in the innovator's mind, and to some degree in actual fact, by producing a stream of innovations. The same, of course, is true for everyone who produces a stream of good work of any kind.

Loyalty and a sense of duty drive inventors to invent as they drive all those capable of loyalty to do their own kinds of work. In time of war the rate of invention jumps because of loyalty to the country. Inventors produce out of loyalty to their organizations and loyalty to individual managers.

Related to aggressiveness are will and determination. A determination to succeed drives the inventor and any associates to success where ability alone does not.

Each of us has competition in some form (see Chap. 14, "Your Competition"). Depending on your personality, competition can inhibit and cause fear or it can stimulate and spur you on. The knowledge that there is a competitor trying to invent a solution to the same problem is a powerful spur to some people.

Some people have a will to fail. It can be an unconscious, self-punishing neurosis which sabotages the efforts of the inventor. I cannot guarantee success to a will to succeed, but I can absolutely guarantee success to a will to fail; failure is a self-fulfilling prophecy.

Persuasiveness. Mastery of the golden art of persuasion is the final attribute of the inventor. It is essential in the real world to lead an invention from a sketch or model to actual commercial use. This is the art of persuasion. The inventor must persuade management or investors to provide a development budget, must persuade the assistants hired or assigned to help to actually help, must persuade marketers to sell the product, and often must personally persuade customers to buy it. If you invent a better mousetrap, the world will leave you severely alone; *you* must beat a path to the *world's* door. (I looked it up. Emerson was not talking about invention; he was talking about workmanship.) The importance of persuasiveness in your life is so great that Chap. 4 is devoted to it.

Aids to inventing

Certain aids to inventing are obvious:

- Sketching to aid visualization
- Clever experimentation
- Skill in approximate calculation

Inventions can be discovered by sensitive observations as well as combined out of modified memories. Patent law speaks of "inventions and discoveries."

Some definable procedures aid invention. These include argument, brainstorming, the application of analogy, and the systematic arraying of elements.

Argument. Two stories illustrate productive arguments:

When I was in the space business in the proposal group of General Dynamics/Astronautics, another engineer, B, who was very bright and an excellent analyst, reacted to any new idea as a mongoose reacts to a snake. I had been appointed proposal leader for Project Prospector of the National Aeronautics and Space Administration (NASA), which was to deposit a series of 2000-lb unmanned payloads on the moon, and it was my job to propose payloads. I conceived a surface vehicle suitable for the deep dust layer which, it was then thought, covered the moon. The vehicle consisted of a 10-ft-diameter hollow sphere driven by the rotation of its interior mechanism, which had an eccentric center of gravity (CG). I presented the scheme to B, who turned red in the face and attacked, as expected. B demonstrated that, with the maximum eccentricity of the CG which could be obtained, the 10-ft sphere would stall against a 20-in rock. No argument; back to the drawing board.

The scheme used a 9-ft flywheel as the energy store. The power came from a solar panel extended through a window on the rolling axis. I added a brake between the flywheel and the sphere, which caused a sharing of angular momentum by flywheel and shell. You can't destroy momentum, so the sphere would keep turning until external impulse (torque × time) absorbed the momentum. When I proposed the new idea, B was reduced to pure invective, which was his standard indicator that the idea would work. (Figure 2.2 shows a 1-ft-diameter working model with its radio control. It would actually skid against a vertical wall in trying to climb up it.)

This example also illustrates alternation between qualitative and quantitative thinking in the inventing process.

A second example of productive argument was the invention of the sorting memory used in the standard Post Office automatic letter-sorting machine. Rabinow Engineering Company had a contract to develop the machine, which had been proposed by Rabinow.

The machine has a traveling chain of letter holders which pass over

Figure 2.2   Roamer moon surface vehicle. (*Courtesy of General Dynamics Corporation.*)

a row of fixed sorting pockets. A setable memory on each carrier recognizes a matching memory on the desired pocket and releases the letter from the holder to drop into the pocket. I was assigned the design of the memories.

It was Rabinow's desire to make the memories purely mechanical because of the customer's limited maintenance capabilities and of his own judgment on reliability based on the performance of an earlier electro-mechanical escort memory he had built for a punched-card machine.

To compare 8-bit mechanical memories a total of several thousand times a second throughout the machine would be very noisy and would wear out a lot of parts, so I was told to try to make the comparison process substantially silent and without impact. This, of course, had to be done on the fly without stopping the chain.

Within a few minutes we were screaming at each other. I said it couldn't be done; he said it had to be done, etc. I finally said that I could prove mathematically that it could not be done. He said, "All right, you have shown that it can't be done, but if you could do it, how would you do it?" I was so upset that this question seemed perfectly reasonable, so I looked up at the ceiling and there, as if projected, was a picture of the device. I sketched it, whereupon he said, "That's it, that's the way we'll do it, but if it was so easy, why did you give me such a hard time?"

The system is shown in Fig. 2.3. Each holder has a long shaft bearing eight code wheels. Each wheel has two axial positions for 0 and 1,

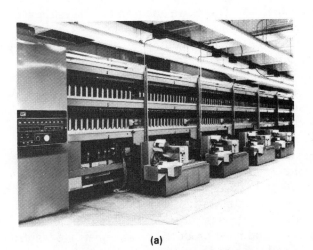

(a)

(b)

Figure 2.3 Letter-sorting machine. (*a*) Overall machine. (*b*) Memory system. (*Courtesy of Rabinow Engineering Co.*)

set to the *holder* binary code. The wheels roll along a flat track. At each pocket the track has eight long, narrow hollows in the eight wheel paths, with each hollow being in either the 0 or the 1 position, set to the *pocket* binary code. If the wheel code and the holder code match, all wheels roll into hollows at the same time and the shaft dips and trips a letter release trapdoor. If even one wheel does not encounter a hollow, that wheel supports the shaft as the other wheels move in air over their hollows. There are thousands of these machines in service today, each with thousands of memories.

Where did this invention come from? The wheels and hollows came from a western comedy movie I remembered in which a horse-drawn carriage jounced its passengers when its wheels ran into potholes in the road. This crazy association was the result of the adrenaline produced by the violent argument. Let's see them program this kind of thing on their computers.

Brainstorming. Brainstorming is a more organized and gentlemanly form of argument. At first the members of the group invent with as little inhibition as they dare and with no criticism permitted aloud. Then the ideas are subjected to criticism to find a survivor. Thus the efforts of a group may surpass those of a single individual. A variation, which I have found most productive, is competitive leapfrogging, in which each member tries to improve on the last idea of another member. The final leap would not be made without the preceding leaps. Much of my present work is such brainstorming for clients. (Ideally, the client's engineers make the last leap so that they will carry the ball with enthusiasm.)

Certain argument practices are helpful, and certain other practices are harmful. When I was just out of engineering school, I argued incessantly with my boss. Finally he lost his temper and bawled the hell out of me. "Kamm, you are just making up arguments to win the argument, not because you think you are right!" He had me dead to rights. I was shocked into silence (which doesn't happen to me often), I apologized, and I have never done that again. (My wife disagrees, so perhaps I just do it less often.) I suggest that you listen to your own arguing style and eliminate arguments which you don't really believe in but which only prevent your having to agree that you were wrong. I call this "argufying."

You should try very hard to keep your temper when arguing. Your arguments will be more rational, and you will be more persuasive.

A valuable by-product of argument, whether passionate or polite, is joint invention. Here the product of the argument is the inextricable contributions of the parties. For example, in the Post Office machine case Rabinow added the mechanism by which my dipping axle released the letter.

Incidentally, if you examine most of the arguments you have had and heard, particularly about human affairs, you will find that most of the arguments on both sides are qualitatively true. The real problem is to establish which truths are quantitatively of greater importance or value than the others; that is very hard. So we just throw the qualitative truths at each other, and no joint conclusion is reached.

Analogs. A formal and disciplined technique of invention is the application of analogs. Analogs are different phenomena subject to the same mathematical equations.

Examples of analogs are:

- Electric fields–magnetic fields
- Electric voltage and current–magnetomotive force (mmf) and flux
- Inertia-inductance
- Stiffness-capacitance (actually 1/capacitance)
- Resistance–viscous drag
- Force-voltage
- Fluid flow–current
- Energy in any form–energy in any other form
- Heat flow–fluid flow–electric current
- Temperature-pressure-voltage
- Waves in water, air, solids–transmission lines–radio

Some analog computers use one twin of an analog to simulate another twin which may be more costly to build and test. In a sense the ultimate analog is nature-mathematics. Digital computers are used to build mathematical models of physical systems.

Analogy is used in inventing by observing a device or phenomenon, remembering a relevant analog, and conceiving an analogous device. In some cases an invention results without a problem to start it. The invention of the magnetic-particle clutch by Rabinow was an application of analogy.

He saw a laboratory demonstration of an electro-rheological (ER) clutch. An ER fluid is one which grows more viscous in an electric field. The fluid used had a suspension of starch particles which became electrostatically attracted into chains when subjected to an electric field. The clutch consisted of a driving disk, a driven disk separated from the driving disk by a small gap, an ER fluid in the gap, and a voltage supply applied between the disks to change the ER fluid viscosity. The clutch worked, but the torque transmitted was very small and the necessary voltage was very high, so it was not a useful device.

(Subsequent improvements in ER fluids may make ER clutches and brakes practical.)

Rabinow reasoned that there should be an electro-magnetic analog using iron particles magnetically attracted into chains between iron disks by an mmf from a surrounding coil. The analog not only worked, but the magnitudes of torque and magnetizing power were useful, so an entirely new and useful device was born, and born in the absence of an immediate need to invent it.

Systematic arraying. There are formal techniques of problem solving which aid the mind in qualitative design. (If they reach the final result by themselves, well and good. There is no law which says that an invention must be better than a result of formal engineering.) Tabulating, on paper, all the elements recalled by free association may help. Researching the literature for more information about those elements you judge to be promising will certainly help. You can design matrices in which different combinations are arrayed, and you can assign numbers representing your value judgments.

Using computers. Some people are trying to do this sort of thing by computer, starting with large databases. They usually deny that they are trying to mechanize "invention" and say that they are only trying to mechanize "design." Their buzzwords are "theory of design" (see Chap. 28). My own belief is that they really hope to make a computer model of the human mind. I wish them luck, but I still encourage you to keep on thinking without fear of technological unemployment in your lifetime.

Note that I am not referring to massive data handling using clerical rules, or to quantitative analysis by equation solving or finite element analysis, or to computer-aided drafting or to computer-aided word processing. For these tasks the digital computer is the most valuable aid device developed in the history of engineering. However, we should not allow ourselves to be misled into believing that computer-aided design (CAD) is design by a computer.

Formal logic versus the real world

The Greek philosopher Aristotle built a system of logic on the basis that all subjects of reasoning fitted defined categories. Then, for example, if all category A is part of category B and if C is in category A, then C is also in category B. Euler circles diagram this logic, and digital computers live on it. We have learned to class things into categories, particularly binary categories: good-bad, big-little, inside-outside, liberal-conservative, science-engineering. Some Asians use the general terms yang-yin and have an artistic design to illustrate it.

The real world isn't that way. Real things differ from each other quantitatively and qualitatively along spectra, and sometimes multi-dimensional spectra, rather than in discrete boxes. The variations may be time-varying, and the things' boundaries may not even be definable. The ultimate time-varying, multiple-spectrum, undefinable-boundary thing is human behavior.

Related to this categorizing is the compulsion to define. An inventor whose mind is constrained to think in sharply defined categories with sharply defined elements won't make it. Even the mathematicians are now talking about "fuzzy sets."

This is not an argument to discard our mathematical and engineering education. It is a warning not to distort your understanding of reality by dividing it into hard boxes, and it is a statement that the process of invention is not an exercise in Aristotelian logic.

Tradeoffs and two-way winners

It is routine engineering practice to trade off the magnitude of one virtue or parameter of a design against that of another to get the best combination. In airplanes one trades off range against payload. In automobiles one trades off power against fuel consumption. There are probably many such duels in your own field, and in many cases there are clusters of parameters to be proportioned. The process of optimizing the parameters has been the subject of mathematic analysis, and there are textbooks on the subject. Often there is no choice, and such optimizing studies must be performed with no alternative in sight.

I suggest that with sufficient effort and ingenuity you may be able to find two-way winners. A two-way winner is a modified or different design which has the advantages of the best in both parameters that must otherwise be traded off. You may be able to devise a change which is both better and cheaper. (In my robot company this was the absolute rule for accepting any proposed change in a design which was already adequate.) Proper selection of proportions and materials may result in a structure which is both stronger and lighter.

Sometimes with a little bit of luck and a great deal of effort a multiple-way winner can be devised. A superb example of this is the container system for ocean shipping. For thousands of years ships were loaded with loose parts, jars, boxes, crates, and barrels. Loading and unloading were laborious, pilferage was common, and damage was frequent. Then a system was devised in which cargo was loaded at the factory or warehouse into large steel containers the size of a trailer truck body. Large cranes were built to handle these containers and to transfer them between ship and shore. Ships were designed to accommodate these containers in neat stacks. The containers themselves were sized to fit onto flatbed truck trailers and onto rail container cars for land transportation. The result is the now almost universal system which has reduced handling costs, pilferage, loading and unloading time, packaging and crating cost, and damage. All of this has been done with a highly cost-efficient technology. The classical approach of optimizing parameters would trade off the comparative cost of security systems versus pilferage loss and the like. The ship container system is a triumph as a multiple-way winner.

In fields other than engineering the search for multiple-way winners is incessant. Such fields include business deals, political policies, and many others where the consequence of a decision is human behavior. All the nonengineering inventors described above are in constant search for two-way and multiple-way winners.

How to benefit from your inventions

On the job. There are many benefits and a few disadvantages to inventing as part of your job. Your prestige goes both up and down, depending on the attitude of each individual to inventors and inventions.

In an organization which is progressive, beyond lip service to "innovation, motherhood, and the flag," your prestige goes up with management and with the other inventors of the organization, and you can compensate—or ignore—your loss of prestige with those others who think of inventors as nuisances and nuts.

In an organization with an antiprogressive bias (regardless of its propaganda) you must balance your unfortunate predilection against your reluctance to change jobs. In such a place you quickly learn to understand the "curse of innovation" and to understand that the way to tell the pioneers from the Indians is that the pioneers are the ones with the arrows in their backs.

Within a single company it is common for different groups to be progressive and antiprogressive. You may be able to transfer within the company without the extreme step of finding another job. Within Convair I was once transferred from a progressive boss to an SOB. I was able to retransfer to the proposal group of the Astronautics Division, where inventing was our business, but I had to threaten to quit before the senior manager would give me permission even to apply to this other division. (I meant it, and he knew I meant it.)

The benefits to inventing on your job can include the following:

- Raises and promotions
- Prestige

- Interesting assignments (e.g., proposal writing)
- Presentation of technical papers (including company-paid travel and meeting other people), publication of articles, and publicity from company PR
- Patents paid for by the company
- Awards (sometimes including real money)
- Job offers by other companies based on the above publicity
- Self-satisfaction (the ego reward described above)
- Job security because of your value

A disadvantage to being an inventor on the job is that you may not be considered stable management material because you are always deviating from the beaten path and therefore may be kept off the management ladder with its different prestige and rewards. Some companies provide a more or less parallel technology ladder, and this provides the raises and promotions mentioned above, but it does not go as high. I was on one, in Convair, and it was real.

However, as the delightfully elegant expression goes, "You'll never get rich as a working stiff." If you are driven to make a lot of money and your strength is in inventing and engineering, what else can you do?

Selling your inventions.   Your first option is to keep your job, invent in noncompetitive fields, and sell your inventions to other manufacturers for the well-known $1 million each.

Forget it.

Companies rarely buy inventions which are not fully developed, in production, and successfully being sold. Why?

- NIH (not-invented-here) factor, a general prejudice. Also a low innovation index. (See below.)
- Most companies have an inventory of undeveloped products which *were* "invented here" but which have been stalled by lack of capital and personnel to get them all the way to market. "Why should we pay you for yours when we already have our own?"
- Fear of legal entanglements, particularly with amateur inventors who may be paranoid and greedy and are presumed to be ignorant of the realities of business costs and risks.

Everything that can happen does happen, so there have been exceptional cases. I have licensed my Decimal Keeper slide rule and zero insertion force connector to companies I had approached for the purpose, but the effort was great and the proceeds were small. Jack Rabinow has done a great deal better. A tiny handful have struck it rich. Try it if you like, but please *do not invest serious personal money to develop an invention for which you do not have a buyer who has actually put up front money! The "buyer" may have a change of mind, leaving you with a serious loss.*

I have sold other patent rights to companies I had contracted to develop products for. This is a real way in which inventors can sell inventions. There are a few free-lance inventors who have established relationships with manufacturers and who are paid royalties for those inventions which are accepted. If you can establish such a relationship, go to it!

There is a path to selling inventions which has a better chance than most. If you invent a product on your job in a field in which you are expert but which does not compete with your employers' products, and if your employers do not choose to exploit it, and if you ask their permission, they may give you the rights to the invention, either free or on a shared-royalty basis. Your invention may be salable because your professional expertise makes it state-of-the-art.

I have won at this twice (my Decimal-Keeper slide rule and my basic modular robot designs), and I have given rights to my own employees. One of them promoted a flexible drafting instrument as a side business, and another invented a vibratory feeder drive which did not shake its base and sold the idea to a feeder company for $10,000 in real money.

My usual advice to inventors who want to develop and sell inventions is to go to the racetrack instead and bet their money on horses. The odds are better, and you get fresh air to boot. However, there is a way:

Entrepreneuring. You start your own business based on your own invention. I have done so four times, with total gains exceeding total losses (there were plenty of both). This is off the subject of inventing itself, so it is described in Chap. 26, "Entrepreneuring."

Obstacles and aids to innovation

If you are a chronic innovator, you will encounter both obstacles and aids to your work from people around you. Unless you are very lucky, you will encounter far more obstacles than aids, and there will be times when you will think of the phrase "the curse of innovation." Here is an outline of what you may encounter.

Most people are hostile to new ideas. Their reasons may include jealousy, recognition of a political threat, emotional conservatism (which is a fancy way of saying that they just feel hostile to new ideas), and the fear of unexpected consequences, which is legitimate conservatism. They will show their hostility in a variety of responses such as angry denunciation, faint praise, active resistance, blank silence, and, in the case of assistants, a will to fail appearing as subtle sabotage.

Your manager, if hostile to innovation, will refrain from praise and rewards for successful innovation, will neglect or reject proposals, and will punish you for errors. (There is no more successful management technique than prevention of error by punishment. Of course, the punishee will thenceforth avoid punishment by not doing anything with any element of risk and by doing as little of anything as possible, but he or she won't make any more errors.)

I have found that manufacturing engineering managers are far worse in this respect than product development and R&D managers. (I except companies like IBM, 3M, and Hewlett-Packard, but more on that subject later.) They assign to manufacturing engineers maintenance tasks and technical clerical tasks which should be handled by technicians, not graduate engineers. They punish errors, discourage innovation, and provide the worst of the available office facilities. As a result, manufacturing engineering has low prestige, is avoided by those who can get better jobs, and is even neglected in engineering schools. All this in spite of the fact that manufacturing technology is as fascinating as any other branch of engineering and offers as much opportunity for innovation as most other branches. This situation bears a major part of the responsibility for United States manufacturing's becoming so noncompetitive with foreign manufacturing that a national disgrace is becoming a national disaster.

My first company, Numerical Control Corp., developed an innovative control system for machine tools. I wrote and telephoned a manufacturing engineer in one of the major Los Angeles aerospace companies and invited him to visit us for a demonstration. He refused to come. He said, "There is no point in my coming because we won't buy it anyway. There might be a maintenance problem after we bought it. If there were, I would be criticized for buying a new machine from a new, small company, and I might be fired. If we buy a less desirable machine from a big company and if there is a problem, no one will criticize me for the selection. Do you think I will risk my job for your damn machine?"

The accounting report practices of American industry include a quarterly report of profit and loss with little segregation of expenditures intended for long-term benefits. The raises, promotions, bonuses, prestige, and job security of senior managers depend on quarterly reports of profit. They are therefore motivated by our crude accounting practices to discourage innovations which do not have immediate return of profit. (See Chap. 8, "Coexisting with Accountants.")

In many large companies senior managers have a relatively short tenure in each of their assignments. Therefore they have little motivation to start spending money and effort on long-term programs whose benefits will be credited to their successors and not to themselves.

Most companies have suggestion boxes for employee proposals. The proposals which are accepted are those which yield immediate profits and not those requiring investment in what might be very much greater long-term profits.

The resistance to innovation of company managers pales before that of venture capital managers and investment bankers. These people boast that their profession is the seeking out of innovation. In fact they seek small increments of innovation in a few fashionable fields. If your innovation is not based on semiconductors, DNA, or coherent light, you will have a tough time promoting investment.

Be of good cheer, all is not dark. There are companies and managers who not only talk about loving innovation but behave and invest as if they mean it. They treat innovators with friendliness, encouragement, praise, tolerance of error, and rewards for success. They read books on how to manage creative people.

How do you find jobs working for such people and avoid jobs working for their opposites? I have only a few suggestions.

Certain industries are more progressive than others. The electronics, pharmaceutical, and chemical industries tend to promote innovations. Certain groups in the aerospace industry (but by no means all) are encouraging. The aerospace industry receives large amounts of government money to encourage and pay for innovation, and it is hard not to seek and accept money. Many government agencies give grants for innovative work to universities and companies. Some R&D groups in universities encourage engineering innovation.

The behavior of individual managers is another story. I joined Convair to find a progressive and capable environment. I prowled in delight this wonderful and diverse high-technology institution and came upon the lofting department. Here drafters in stockinged feet walked on large aluminum sheets and drew the contours of airplanes by using magnifying glasses, scribers, and steel scales.

I had done some pioneer work in the numerical control of machine tools, so I proposed to the department head that we make a numerically controlled (NC) lofting machine (essentially a large, lightly loaded NC milling machine) which would be faster and more accurate than manual lofting and would drastically reduce the labor cost. He glared at me and demanded, "Are you trying to take the bread out of our mouths?" The answer was yes, so I went away. (Now, of course, NC lofting is standard.)

The book *In Search of Excellence*[33] is a study of companies which actively promote innovation. I recommend it as an aid in your search.

In general the metalworking and machine tool industries are reluctant to invest in innovation, so major parts of our machine tool and automobile orders go overseas.

Perhaps the worst industry I know is the women's shoe industry in the United States. I don't know how much of it even survives today; most women's shoes are now imported. Some years ago the Department of Commerce held an elaborate conference to save the shoe industry. I was invited as one of a panel of inventors to contribute innovations (aside from myself it was an august group). Executives of shoe-manufacturing companies, marketers, and an executive of the leading shoe machinery company attended.

The marketer told how she could not get new styles accepted until they appeared from abroad in the stores. The shoe machinery executive said that it had become typical that his company would develop a new machine and demonstrate it to a shoe company executive who would love it. The shoe executive would then ask:

"Who is already using the new machine so I can see it in commercial service before I buy it?"

"No one. It's brand-new. You have the opportunity to be the first to benefit from its advantages."

"Oh. Well, I'll buy the second machine after the first has established a successful track record."

So the machinery company stopped developing new shoe machines and did its further growth in electronic assembly machines.

The shoe manufacturers spoke a single theme. The solution of their problem was protective tariffs.

Innovation index

In dealing with individuals I have developed an innovation index (II) which I use to classify each with respect to receptivity to innovation. Since I cannot ask customers and clients to take a test, I listen for key phrases and behaviors. You may find it amusing and perhaps useful in allocating your own promotional efforts:

II 0. Attitude based on benefits or costs to the listener, not his or her attitude toward merit

II 1. Very hostile

- Anger, rage (e.g., engineer B, above).

- Instant "No!"

II 2. Somewhat hostile

- "It won't work."
- "It's not practical."
- "It's too expensive."
- "Authority X disagrees."
- "The trouble with that is..." pause (while he or she thinks up an objection).
- "It's too late to change."
- "But we have always done it the present way."
- "I think someone has already done that."
- Evasion or silence, without spoken criticism.

II 3. Neutral, objective

- "Well..., maybe...."
- "Let me think about this."
- "Prove it."
- Questions.
- Analysis.

II 4. Friendly

- "I hope you're right."
- "Congratulations."

II 5. Enthusiastic

- "Wonderful!"
- Joy and enthusiasm.

What is your own innovation index? If you have never asked yourself before, you probably don't know. I suggest that you watch your own reaction to other people's suggestions and new ideas over the next few weeks. If you are displeased with your own performance, don't tell anyone; just try to improve it. I have learned to accept advice and ideas and have benefited greatly by doing so. I believe I am an II 3 to 4, but I know I sometimes backslide to a 2.

As a suggestion to other inventors, we are all II 5 to our own new ideas. It may be a good idea to let these ideas cool off for a day or two before exposing them to others.

I recently returned from a tour of Hungary, the most liberal of the iron curtain countries. If you think the governing II in the United States is bad, you haven't seen anything!

Now let me close with my favorite inventor-entrepreneur story.

Nathan Zepell was a European graduate engineer who kept himself off the death list in a German concentration camp during World War II by making toys for the camp commander's children. (I heard him tell the story in detail.) After the war he came to New York, where he invented a pen pocket clip which triggered the retraction spring when the pen was put into a pocket. He tried persistently to sell the invention to the C Pen Co. with the usual result. He finally promised that if the company would send him a letter explaining its rejection, he would not bother it again. The company did, with 10 reasons ("impractical," "unnecessary," "too expensive," etc.—the usual stuff). Nate kept his promise.

He then found a small manufacturer of souvenir pens who liked the idea. Together they built tools and started to make pens which included the invention.

One of these pens landed on the desk of Mr. C, president of the C Pen Co. Motion-picture climax. Invitation to meeting of board of directors. Red carpet. Letter brushed aside. Contract. Royalties.

A few years later I visited Nate in his beautiful home on a mountaintop in Santa Barbara, where he had one large room fitted out as a laboratory and twin white Lincoln Continentals for him and his wife parked outside.

Everything that can happen does happen.

REFERENCES

Publications available from the U.S. Patent and Trademark Office

Orders should be addressed to U.S. Department of Commerce, Patent and Trademark Office, Washington, D.C. 20231. Remittances should be made payable to Commissioner of Patents and Trademarks. Postage stamps, Superintendent of Documents coupons, or other government coupons are not acceptable in payment of Patent and Trademark Office fees. Prices are for fall, 1987.

1. Patents. Copies of the specification and drawings of all patents are available at $1.50 each. When ordering, identify the patent by the patent number.
2. Trademarks. Printed copies of registered marks are sold at $1.00 each. Printed copy of trademark showing title and/or status, $6.50; for certifying trademark records, $3.50 per certificate; and for photocopies or other reproductions of records, drawings, or printed material, $0.30 per page of the material copied. For other items and services that may be furnished for which fees are not specified, such charges as may be determined by the commissioner will be at actual cost. When ordering, identify

the trademark by registration number or the name of the registrant and the approximate date of registration.

3. *Q & A about Patents.* Brief, nontechnical answers to questions most frequently asked about patents; revised June 1986. Free.
4. *Q & A about Plant Patents.* Same as Ref. 3 for plant patents; n.d. Free.
5. *Q & A about Trademarks.* Same as Ref. 3 for trademarks; n.d. Free.

Publications available from Superintendent of Documents

Orders should be addressed and remittances made payable to the Superintendent of Documents, U.S. Government Printing Office, Washington, D.C. 20402. Prices are for fall, 1987.

6. Annual indexes. An index of the patents issued each year is published in two volumes, one an alphabetical index of patentees and the other an index by subject matter of inventions. The two parts are sold separately. Price varies from year to year, depending upon the size of the publication. An annual index of trademarks contains an alphabetical index of trademark registrants, registration numbers, dates published, classification of goods for which registered, and decisions published during the calendar year. Price varies from year to year, depending upon size of the publication.
7. *Annual Report, Fiscal Year 1986, Commissioner of Patents and Trademarks.*
8. *General Information Concerning Patents.* Contains a vast amount of general information concerning the application for and granting of patents, expressed in nontechnical language for the layperson. Revised Oct. 1, 1986. Copies available from the Superintendent of Documents, $3.00 each. 003-004-00583-1.
9. *General Information Concerning Trademarks.* Same as Ref. 8, for trademarks; n.d. Copies available from the Superintendent of Documents, $3.25 each. 003-004-00582-3.
10. *Guide for Patent Draftsmen.* Patent and Trademark Office requirements for patent drawings with illustrations; n.d. Price, $2.25. 003-004-00570-0.
11. *Manual of Classification (POM).* Loose-leaf volume listing the numbers and descriptive titles of the more than 300 classes and 66,000 subclasses used in the subject classification. Substitute and additional pages, which are included in the subscription service, are issued from time to time; n.d. Subscription, $65.00 ($16.25 additional for foreign mailing).
12. *Manual of Patent Examining Procedure.* Loose-leaf manual which serves primarily as a detailed reference work on patent-examining practice and procedure for the Patent Examining Corps. Subscription service includes basic manual, quarterly revisions, and change notices; n.d. Subscription, $70.00 ($17.50 additional for foreign mailing). 003-004-81002-5.
13. *Patent and Trademark Forms Booklet.* Designed for patent and trademark applicants, attorneys, and agents, this booklet contains full-size reproducible copies of frequently used patent and trademark application forms and related forms; forms in English and other languages; n.d. Price, $15.00. 003-004-00569-6.
14. *Patent Attorneys and Agents Registered to Practice before the U.S. Patent Office.* An alphabetically and a geographically arranged listing of patent attorneys and agents registered to practice before the U.S. Patent and Trademark Office; n.d. Price, $9.00. 003-004-00573-4.
15. *Patent Official Gazette.* The official journal of the Patent and Trademark Office relating to patents. Issued each Tuesday, simultaneously with the weekly issuance of patents, it contains a selected figure of the drawings and an abstract of each patent granted, indexes of patents, list of patents available for license or sale, and general information such as orders, notices, changes in rules, and changes in classification. Annual subscription, $250.00, domestic; $360.00, first class ($312.00 for foreign mailing). Single copy, $7.50, domestic.
16. *Patents and Inventions—An Information Aid to Inventors.* The purpose of this pub-

lication is to aid inventors in deciding whether to apply to patents in obtaining patent protection and promoting their inventions; n.d. Price, $3.25. 003-004-00545-9.

17. *Rules of Practice in Patent Cases.* A consolidation of patent rules and a revised index; n.d. Price, $5.00. 003-004-00597-1.

18. *The Story of the United States Patent and Trademark Office.* July 1981. Price, $4.75. 003-004-00579-3.

19. *37 Code of Federal Regulations.* Available from the Superintendent of Documents; n.d. Price, $6.50. 022-003-94174-6.

20. *Trademark Manual of Examining Procedures (TMEP).* n.d. $23.00 ($5.75 additional for foreign mailing).

21. *Trademark Official Gazette.* The official journal of the Patent and Trademark Office relating to trademarks. Published every Tuesday, it contains an illustration of each trademark published for opposition, a list of trademarks registered, a classified list of registered trademarks, and Patent Office notices. Annual subscription, $205.00 ($256.25 for foreign mailing). Single copy, $5.00, domestic; $6.25, foreign.

22. *Trademark Rules of Practice of the Patent Office with Forms and Statutes (TRPP).* Contains the rules and forms prescribed by the commissioner of patents and trademarks for the registration of trademarks and a compilation of trademark laws in force; n.d. Price, $18.00 ($22.50 for foreign mailing).

Books on inventions, patents, and creativity

23. Amernick, B.: *Patent Law for Nonlawyers: A Guide for the Engineer, Technologist, and Manager,* 1st ed., Van Nostrand Reinhold, New York, 1986.

24. Brown, A. E.: "Invention and Innovation—Who and How," *Chemtech,* December 1973. Comparison between invention and implementation, particularly in the chemical industry.

25. Cook, C. L.: *Inventor's Guide,* 1st ed., C. L. Cook, P.O. Box 1511, Slidell, La. 70458, 1981.

26. Feinman, R. P.: *Surely You're Joking, Mr. Feinman!* 1st ed., Bantam, New York, 1986.

27. Jewkes, J., D. Sowers, and R. Stillerman: *The Sources of Invention,* Macmillan, London, 1958. Out of print.

28. Konold, W. G., et al.: *What Every Engineer Should Know about Patents,* 1st ed., Marcel Dekker, New York, 1979.

29. Middendorf, W. H.: *What Every Engineer Should Know about Inventing,* 1st ed., Marcel Dekker, New York, 1981.

30. Mogavero, L. N., and R. S. Shane: *What Every Engineer Should Know about Technology Transfer and Innovation,* 1st ed., Marcel Dekker, New York, 1982.

31. Neumeyer, F., and J. Stedman: *The Employed Inventor in the United States,* 1st ed., M.I.T., Cambridge, Mass., 1971. A scholarly study of the law and specific company, university, and government agency policies.

32. Norris, K.: *The Inventor's Guide to Low-Cost Patenting: How to Patent Your Own Invention and Save Hundreds,* 1st ed., Macmillan, New York, 1985.

33. Peters, T. J., and R. H. Waterman, Jr.: *In Search of Excellence: Lessons from America's Best Run Companies,* 1st ed., Harper & Row, New York, 1982.

34. Sandler, Ben-Zion: *Creative Machine Design: Design Innovations and the Right Solutions,* 1st ed., Paragon House, New York, 1986.

35. Sobel, R., and D. Sicilia: *The Entrepreneurs: An American Adventure,* 1st ed., Houghton Mifflin, Boston, 1986.

36. Tuska, C.: *Inventors and Inventions,* McGraw-Hill, New York, 1957. Data, inventor stories, large bibliography; out of print.

37. Wilson, M. A.: *American Science and Invention,* Simon & Schuster, New York, 1954. Inventor stories; out of print.

People Problems

3

The Politics of Design

What Is Politics?

Power to effect one's will is what politics is about. People seek power for the pleasure of having it, for prestige, for the fear it induces in others, and for the benefits it helps to bring. There are many kinds of power:

- Magic power is the imagined ability to cause changes in the real world by manipulating symbols and by incantations.
- Design power is the ability to cause the product to perform in accordance with your will.
- Political power is the ability to cause people to perform in accordance with your will.

The Random House Dictionary of the English Language definition of politics is "use of intrigue or strategy in obtaining a position of power or control, as in business, universities, government, etc."

The connotation of immorality is common but not universal. The Founding Fathers, all our presidents, both good and bad, all business entrepreneurs (including me), and many great engineers and scientists (e.g., Joseph Strauss of the Golden Gate Bridge, Edison, Tesla, and Westinghouse) have all been politicians. There is a complete moral spectrum of people who seek power, from the worst to the best.

There is no carrying of concept to product without politics. The better you are as a politician, the more successful you will be in getting your designs to market and in having them get there as you want them designed. You are a politician whether you like it or not, so be a successful one.

Politics and You

The people you relate to politically include:

- Your management hierarchy
- Your coworkers
- Your assistants
- Other departments: manufacturing (products and processes), marketing, accounting, services
- Customers
- Secretaries

Some of your political problems as a design engineer are:

- Your workplace
- Your budget
- Your assignments
- Your supervisors
- Your coworkers
- Your company's other departments
- Your customers
- Your vendors

Political Processes

Virtuous political processes which you may practice to cause people to perform as you want them to include:

- Doing good work
- Making friends (networking, contacts)
- Verbal persuasion (see Chap. 4)
- Written persuasion (see Chap. 4)
- Favors
- Trading benefits
- Flattery
- Small gifts (e.g., flowers to secretaries, greeting cards)

- Entertainment

Vicious political processes which you may practice or which may be practiced on you include:

- Doing bad work
- Making enemies
- Favoritism
- Evasion
- Distortion
- Threats
- Conspiracy
- Treachery
- Bribery
- Lies
- Fraud

These lists are not just theoretical. In my own career I have been on both the acting and the receiving ends of every item on the list except the last six, and I have been on the victim end of each of the last six. (Yes, I have done good work and I have done bad work. Mostly good.)

Rank and status are associated with political power and are among the motives for seeking political power. Rank and status not only feel good but display and reinforce that power. Among the components of rank and status, other than cash income and a piece of the action, are:

- *Titles*

- *Office facilities.* Windows, rugs, large area, quality furniture and decorations, secretary (or secretaries), location (mahogany row), private or semiprivate bathroom.

- *Badges and ID cards.* Identification of class of employment, rank, duration of employment, and awards. In my aerospace company yellow meant "hourly," candy stripes meant "salaried," red meant "supervisory" or "high-ranking salaried," and blue meant "corporate staff."

- *Dress codes.* Usually unofficial; too complicated to detail here.

- *Perks.* Long vacations, club memberships, favorable sick leave and time-off rules, company car, executive dining room, even an executive barber shop.

- *Designation as "staff." The in group.*

The military go the limit in all these classifications.

In some companies, such as in computer software, there is an inverse show. The president wears a checked shirt with no tie and drives an old, battered car. Scraggly beards and sandals without socks indicate status as a professional.

In our personal lives the desire for status shows in our cars, license plates, clothes, jewelry, memberships, and many other things.

In the United States the desire to exhibit status is limited and inhibited by our democratic tradition. In Germany a Ph.D. who has taught in college and is now head of a company is addressed as Herr Doktor Professor Director Schmidt. In England nothing matches the award of a title of nobility (knight, dame, lord, etc.), which is specifically forbidden by the United States Constitution. We have only "The Honorable Mr. Smith" instead. We do permit the exhibition of some professional and academic titles on stationery (for example, Ph.D., Lic. P.E., C.P.A., M.D., D.D.S.) and in conversation [for example, Your Honor, Professor, Doctor (for Ph.D. or M.D.), Counselor]. Job titles are almost unlimited and are permitted on business cards and under your signature in letters.

How to Politick

Observe and analyze the political processes around you in the world, nation, company, and family. Your deeper understanding and better practice will contribute mightily to your success.

Don't hide your light under a bushel and wait for the world to discover how great you are. All successful entrepreneurs and inventors have been publicity hounds, including me. The major reason for publishing technical papers is the personal publicity they bring.

Don't strut and boast, because the effect will be bad. Distribute copies of your reports and papers; meet the editors of your house organ and trade journals and everyone else. Give papers at technical society meetings, and meet the other visitors to those meetings. Write letters to the editor. Refer to your achievements in conversation, but casually. In short, boast while seeming modest.

Use the political processes described above. It's up to you which ones you choose to avoid on ethical grounds.

Always remember that being a good politician is an essential requirement for success.

Persuasion: The Golden Art

In politics, coercion is the stick; persuasion is the carrot. Selling is a special case of persuasion, but the word "sell" is often used to mean "persuade."

Why and Who

Why must you make an active effort to persuade other people to think and act as you wish? Should not the technical merits of your engineering speak sufficiently for themselves without having to pull a lot of nonengineering psychology and language skills into the act?

In the first place, the technical merits of your engineering designs are not absolute. No one will question your conclusion that a current is 17.5 A or that a stress is 14,300 lb/in^2. However, suppose that you choose to couple a lever to a distant load by a mechanical linkage and that a first technical competitor (who may be your boss) says that an electro-mechanical servomechanism would be better and a second technical competitor says that a sealed hydraulic coupling would be best? There are many judgmental (i.e., nonmathematical) considerations in making the decision, including reliability, maintenance, preferred manufacturing processes, management and customer prejudices, and so on. If that mechanical linkage is important to you, you had better marshal a persuasive case that it is the best, or one of the competing systems will prevail.

In the second place, there are conflicting motives other than the desire to choose the technically best. Your competitor may seek a decision which allocates budget to another department. Your marketing department may seek a decision which aids in an immediate sale or contributes to its long-range sales hopes. Your manufacturing managers may resist adopting a technology new to them, with consequent problems of changing labor skills and overcoming their own conservatism and insecurity feelings. Your engineering management is concerned with maintaining a steady backlog for the different portions of its staff. Your customer, if you are negotiating the features of the product, has a complex of motives different from yours, and sometimes your "customer" is actually several people from different departments, all with different wishes. This is the real world in which you must work.

In the third place, there is human entropy. Human activity tends to diffuse and to peter out unless the humans involved are continually persuaded to toe the line set for them. (I speak of actual work, not of time card charges.) I know about the formal authorities and responsibilities of job descriptions and organization charts. But if you want something done, keep an eye on it and persuade the people to do what they have been instructed to do and are paid to do but may or may not be doing with goodwill or at all.

Always remember that people believe what they want to believe, often for reasons of which they are not really conscious, regardless of where objective logic may lead. They may justify their beliefs with all the techniques of prejudiced journalism: selection of facts, selective emphasis, distortion, choice of words which imply good or bad, errors, and outright lies. You must persuade them to believe what you want them to believe if you wish to be successful.

On whom must you practice the nonacademic engineering arts of persuasion? On everybody who can affect the results you wish.

Unless you are working on a one-person project or detail assignment, your work must mesh with the work of other professionals whose rank is close to your own. You should cooperate with them, and you should persuade them to cooperate with you. I know they are paid to do so, but to varying degrees, depending on the personalities involved, there will be either persuasion or neglect and conflict.

You have or will have assistants assigned to help you. These include other engineers, drafters, laboratory technicians, model makers, vendors, and secretaries. They have been assigned and are paid to follow your instructions, to help you, and to abide by your decisions. To one degree or another they will actually do so. But in the real world you will find that they do so faster and with less resistance, evasion, and misunderstanding if you also persuade them to want to do so.

You yourself are an assistant to your own managers or bosses. You like or dislike many of the things they tell you (or do not tell you) and many of the instructions and authorizations they give to you (or do not give to you). In fact, you may or may not like them or different aspects of them. You will be more successful if you persuade them to tell you and instruct you what you wish to be told and instructed.

Sooner or later you will deal with your company's marketers. You can and should learn from them a tremendous amount about the world outside the company which will influence your designs. Conversely you will have product ideas which are new to them, and you probably have a greater depth of engineering knowledge than they do. You have an opportunity to succeed in promoting your own ideas and yourself if you persuade them to accept and advocate your ideas.

Your factory management and its manufacturing engineers may have a strong influence on what your top management decides to make or not make and certainly on how it is made. They, too, are useful targets of your persuasiveness.

Your customers are the ultimate persuadees. As sales types like to say, "Nothing happens until someone sells something to someone." Your own contact with customers may be through your marketing department, may be in the form of proposals written by you and requested and delivered by your marketers, and, if you are lucky, may be directly face to face with your customer. The ultimate purpose of marketing and sales people is to persuade customers to issue purchase orders.

Customer persuasion can be divided into solicited and unsolicited persuasion. Solicited persuasion occurs when a customer asks you to propose or quote on a product which the customer specifies. The specification is more or less detailed, and asking you is usually preceded by missionary work by your marketers to persuade the customer that there may be benefits in asking you. The federal government advertises solicited requests for proposal or quotation in its *Commerce Daily* and sometimes awards contracts to companies which respond with low bids but which they have never heard of before.

Unsolicited customer persuasion occurs when your company develops a new product and then persuades customers to buy it. It also occurs when you submit an unsolicited proposal to a customer (often a government agency) to perform a piece of research or to develop a new product designed specially for its use.

I will not extend this recital of who and why to persuade into your personal life but will leave that as an exercise for the student. I, too, have been a bachelor and am now a husband and a father.

If I have persuaded you that you must persuade others and who these others are, I should at least try to explain the how of the Golden Art.

How

Young engineers fresh from school believe that if they do good work its merit will persuade others to use it. They are absolutely right. The most persuasive of all the arguments in selling engineering ideas and work is their intrinsic value. The remainder of this chapter deals with situations and techniques in which self-evident merit alone is insufficient to overcome the resistance of the persuadee.

My favorite story of selling by merit was the space vehicle simulator my first company sold to my former aerospace employer. My old boss telephoned me and ordered a half day of consulting. I sent John Toth, who knew more about the subject than I did. At the end of the afternoon the aerospace people asked him to return for a second half day because he had given them so many new ideas. The scope of the work grew and grew. When they had hardware-building problems, we built hardware. When the two hardware programs merged, we invited their employees to work in our shop at no charge. At the end of the program our share was 750,000 1964 dollars, up in small increments merited by continual good performance, starting from the original half day.

When you persuade, use 100 percent facts and 0 percent snow. Read Ref. 6, "How to Write Potent Copy," by Ogilvy, and do exactly what he advises. People are so bombarded by advertising that they are trained and sensitized to any phony sales pitch. You can put a pretty picture on the cover of an engineering proposal; it will attract attention, but it will reduce and not enhance the effectiveness of the proposal. On the other hand, good typography, plenty of appropriate graphics, good paper, and, above all, good English will lend credibility to the proposal.

Make your own behavior more likable. People are more willing to be persuaded by those they like than by those they dislike. Observe your behavior. Do you irritate people or please them? Do you express interest in the affairs of others? Are you sarcastic, discourteous, offensive to the feelings of others? (I said "feelings," not "thoughts.") Do you dress in a conventional way? (Sorry, but it can help. Note that I said "dress," not "think." See Refs. 4 and 5.) Do you normally scowl or smile? Do you "argufy"? (This is my word for arguing a lot just to win, not because the argument is valid and useful.) This paragraph is not intended to be an instant psychoanalysis. It does not inquire into why you might do any of the antipersuasive things. But it points out harmful practices which you can become aware of and, by direct self-discipline, change to increase your professional success.

There are two sentences which instantly soothe your adversary and increase your own prestige at the same time: "I don't know," and "I was wrong." These are not just confessions of human fallibility but are demonstrations of your own integrity. It should be a matter of professional pride not to fake knowledge you don't have and not to stonewall a position in which you no longer believe. In return your associates will think of you as nonphony, the reverse of the tag often pinned on the unconventional.

In communicating to people not trained in engineering, use analogy. Everyone has some understanding of cars, airplanes, household appliances, and the statics and dynamics of their own bodies. Each has special knowledge, some of which you share (e.g., manufacturing technology with manufacturing people). You can explain the principles of your ideas, perhaps not the details or the mathematics, by comparison with the knowledge already shared. I once gave a talk entitled "Automation in Your Life" to a meeting of the American Association of University Women. All the women were very bright and had a liberal arts college education, but few had studied more than elementary science or mathematics. Yet before the evening was over they had learned an insight into the workings of cars, appliances, and manufacturing automation. They really understood what was meant by "closed-loop feedback control," "open-loop control," "transducers," "amplifiers," and "actuators," and they stayed awake and enjoyed the whole thing.

Show that your design benefits the persuadee. We are all motivated largely by what we believe is our own self-interest regardless of our instructions, salary, and duty. If you demonstrate that what you want is in the interest of the persuadee, you will tend to persuade him or her to want it too.

Appeal to the emotions of the persuadee. Show that your design benefits the persuadee's self, group, company, or country and associate yourself with that group, company, or country. This is easier said than done, but the great leaders of the world became great by doing just this.

All people are different, so suit your persuasion to the individual responses of the person you are persuading.

If you want something desperately, hide the fact and appear casual; people are sometimes perverse.

There is negative persuasion which uses this effect. For example, one of the chronic problems in managing a factory is to persuade workers to use safety equipment such as safety glasses. After a period of frustration from this problem I called a general meeting and said: "From now on I will not pester you to wear safety glasses. You provide the eyes, and the company will provide the insurance. End of meeting." That very afternoon one of the men came into my office, shaking. In his hand was a pair of safety glasses with a star-shaped crack pattern right off the poster. "Thanks, Larry," he said, "Thanks again."

Compete for repeated attention. You will rarely change someone's mind permanently in one interview. Even if you do, the person is likely to backslide after you have parted. Find excuses to repeat your case. The successful salesperson makes repeated visits to the customer, each time devising some excuse to prevent the appearance of harassment. This is called "following up." When I sold my robots, I kept following up and always had a new piece of information or a revised proposal to interest and missionary my prospective customer.

If you want an organization (your own company, your customer's company, a government body) to do something, do *not* find the "key" person and persuade only that person. Sell to everyone who can influence the decision, and sell to all of them at the same time.

When I joined the proposal group at General Dynamics/Astronautics, I heard a lecture by our chief Washington salesman. He said: "Everyone knows that the contract issues when the general signs a requisition, so forget the others, get to the general, and sell the general. Right? Wrong! The general has long since forgotten the detailed technology and is largely concerned with political effects, so he won't sign until he is assured that the colonel has approved. Similarly with the colonel and the major, etc. Therefore you must sell the lieutenant *and* the captain *and* the major *and* the colonel *and* the general, and *then* the general will sign!"

Listen as well as talk. Not only will the persuadee like you for providing a respectful audience but he or she may tell you what is really wanted so that you can adapt your case to those wants. In salesperson's slang, you will learn the persuadee's "hot buttons."

Avoid "permutation" meetings. Permutation meetings are meetings of five people held four at a time, so no final decisions can be made. (Yes, mathematically these are "combination" meetings rather than permutation meetings but permutation sounds more high-tech.)

Bootleg your proposal partway, and exhibit successful progress. The Sidewinder missile was bootlegged in defiance of Navy policy, but its technical success sold the design. But remember that you are playing with fire. If your early results are not successful and if you have an unsympathetic management, this is a dandy way to get fired.

Use acronyms, trade names and trademarks, and slogans (e.g., radar, zero defects, know-how) and descriptive phrases with favorable connotations (my robots really did have both "passive homing guidance" and "active homing guidance"). The wordmanship adds absolutely nothing to the technical merits of your design, but it puts favorable feelings in the minds of the persuadees and thus encourages them to accept your design.

Leadership

Leadership is the persuasion of equals and subordinates. Everything above applies, but there is a lot more to it.

It is conventional wisdom that leadership is persuasion done only by encouragement and never by coercion. The successful leaders who

preach this sermon really mean it, but unconsciously they do not practice it. It took me 25 years of struggling to follow the rule to realize
that the other half of the formula is the iron fist in the velvet glove.

For example, I was told this conventional wisdom with absolute sincerity by the former general in command of Sandhurst, the British
military academy, who knew perfectly well what happened to a soldier
who had disobeyed orders and who knew that the soldier knew it too.

I have recognized late in life, in encouragement-type leaders, a
quiet edge to their voices and words which suggest the unpleasantness
consequent to not being persuaded.

Discipline by reward and punishment is an essential part of leadership. If and when you are a supervisor, you can reward your subordinates for obedience and good performance by:

- Raises
- Promotions
- Praise
- Favorable assignments
- Job security

You can punish by the reverse of each of these. You should practice
all of them and not just promise or threaten, or your subordinates will
learn to ignore you.

To some the tone and manner of authority come naturally; to others
they can be learned if the lessons are taught. I realized very late in life
that the first lesson in the first management course in my aerospace
company was not just a joke. The lesson was: "It helps, in being a manager, to be 6 feet tall and have a loud, deep voice." Napoleon did pretty
well without the height qualification, but he made up for it in other
ways.

There are inspiring leaders who successfully persuade by example,
and I have followed one. You should try, as I have tried, but they are
hard acts to follow. I hope you have the talent.

Just as inventiveness requires a combination of talent and effort, so
does persuasiveness. I know a teenager, quite bright, who refuses to
do homework but gets by in school by sheer charm. He has a lovable,
persuasive personality which gets him what he wants without work.
He would consider this chapter completely obvious. If he can overcome
the ethical examples set by his parents, he will be a great con man.

You use speech and writing to persuade. You speak to individuals,
small groups, and large groups. Speech is a learnable art, and you
should learn it. Speak loud enough to be heard, pronounce clearly,

omit all but essentials the first time around, use specific words rather than general words, use complete sentences including verbs, and try to sound like Winston Churchill. (Easy for me to say.) See Ref. 2.

Join Toastmasters. Toastmasters is an organization which holds meetings at which the members give talks to each other and teach each other how to do it better.

Learn to use slides and viewgraphs. If you work in a big company, it has an art department which will make you slides and viewgraphs and help you to design them. Use humor. (Be careful to avoid old jokes!) Use quotations.

After you have written something, let it cool off for a day and then read it. You will be surprised by how the meaning can change overnight. After you have corrected the piece, show it to a friend. It is most discouraging to find that what is crystal-clear to yourself can be so badly misunderstood by someone you used to think was bright. The bad news is that you must not argue or explain. If the friend misunderstands, so will your intended readers, who will not have the benefit of your arguments and explanations. *You* must change what you have written. Good writing is like good design development: you must repeatedly refine your work.

Why I Learned to Write

Writing is the bane of engineers; we are almost proud to do it badly. Yet we must do some of our persuading in letters, reports, and proposals.

I got lucky. I told you how I transferred to the proposal department of Convair/Astronautics, where we invented for a living. What I did not say, and did not know until I got there, was that our work came out in the form of written proposals to customers. A proposal won if it persuaded the customer to buy or lost if it did not.

For the first time in my life I was in a group of very fine engineers whom I respected enormously but who actually kept talking about sentence structure and choice of words. Did I change fast!

This book is an example of how I think engineers should write. I will not give you the rules I follow because I want you to get the references at the end of this chapter and read them. They are short, easy to read, and worth their weight in platinum. (See Refs. 3, 6, and 7.)

The need for persuasion occurs when there is conflict. Whose will shall prevail? One approach is to negotiate a compromise. Most people will be persuaded by a proposed reasonable compromise to abandon a stonewall position rather than wage a prolonged battle. There are books and courses on negotiation, and you may do well to read or take one. A second approach to conflict is to invent a two-way winner (remember the discussion in Chap. 2) in which both parties leave happy.

My last suggestion is to read Dale Carnegie. His book[1] and his courses have been helping people persuade other people for over 40 years.

REFERENCES

1. Carnegie, Dale: *How to Win Friends and Influence People,* rev. ed., Pocket Books, a division of Simon & Schuster, New York, 1983.
2. Flesch, R.: *The Art of Plain Talk,* rev. ed., Macmillan, New York, 1985.
3. ———: *The Art of Readable Writing,* rev. ed., Macmillan, New York, 1986.
4. Molloy, John T.: *Dress for Success,* 1st ed., Warner Books, New York, 1976.
5. ———: *The Woman's Dress for Success Book,* 1st ed., Warner Books, New York, 1978.
6. Ogilvy, D.: "How to Write Potent Copy," *Confessions of an Advertising Man,* Atheneum, New York, 1st ed., 1963, 11th printing, 1986, chap. VI.
7. Strunk, W., and E. White: *The Elements of Style,* 3d ed., Macmillan, New York, 1979.
8. Tenner, Edward: *TechSpeak,* 1st ed., Crown, New York, 1986.

5

Your Ethics

A father brought his son into his retail business. He said, "First I will give you a lesson in ethics. Suppose a customer gives me a 5-dollar bill and I give him change. After he leaves, I discover that it is a 10-dollar bill. We now have a problem in ethics: Should I or should I not tell my partner?"

Morals

It is *not* my purpose to exhort you to be either honest or dishonest; that is a matter between you and your conscience. My subject is success, and there have been many successful and unsuccessful crooks, and there have been many successful and unsuccessful honest people.

Twenty-one ethical problems you will have

It *is* my purpose to present you with a series of ethical problems which may arise in your life so that you will have thought of them before they arise and will not have to make snap judgments when you face them. I have no absolute answers for you to these problems, but it will help you to face them before they arise.

1. We all fill out time cards so that the cost of our time will be charged to the proper accounts. Sometimes we or our managers are embarrassed because a job is taking longer than expected and are tempted to charge time to a different job or account to mask the overrun. This is not stealing, after all, because no money changes hands; it

just misleads the accountants, who deserve no better anyway. Two ethical questions: Should you do this on your own? What should you do if you manager asks you to do it or if you learn that your manager is changing your time card?

Sometimes money might change hands. An engineer came to me in my first company and asked if it was OK to make such a transfer. The engineer obviously intended only to make things look better. However, I knew that the reduced account was a fixed-price job and that we would make or lose the same amount regardless of the accounting, but the increased account was a time-and-material contract and the increased charge would put free money into our pocket; the engineer had handed me a license to steal. Even if we were caught, we could plead honest error and would get away with it. After a few seconds of shock I decided that either I would be a crook or I would not, so I sent the engineer away with my thanks for such zeal and an instruction to play it straight. I still feel pretty smug about the episode.

2. Is it ethical to ask for a raise, or should you wait for your employer's scheduled review? Is your request a form of blackmail, an implied threat that you will leave if a raise does not come through?

3. Is it ethical to take, for your own use, pencils, paper, and other items of very small cost because of the time it would take to go out and buy them at retail? If it is quixotic to make an issue of one paper clip, where do you draw the line?

To extend this problem, is it ethical or at least acceptable to take objects of large value if you judge that you can get away with taking them? How about synthetic transactions (i.e., fraud)?

4. Is it ethical to accept a cheap lunch from a salesperson whose visit extends through the noon hour? How about an expensive lunch? Drinks after work? An expensive dinner? Tickets to a ball game? A weekend on the boss's yacht? (Don't laugh—your pronounced judgment of the product may influence very large sums of money.) The gift of a souvenir ballpoint pen with a company name and worth less than a dollar? Gifts valued at $1, $10, $10,000? All the above have figured in real transactions. (On the other hand, I remember making a sales call to a government laboratory after a major scandal. My host just laughed when I tried to pay for the host's 85-cent lunch in the lab's cafeteria, saying "No way!")

5. What about using ideas generated by someone else, either inside or outside your organization, without payment of money, credit, or opportunity? When is such an idea to be considered "intellectual property," and when is it to be considered "in the public domain"? What ground rules apply to intellectual property which you judge to be safe to take without legal consequences?

A salesperson discloses a company's unpublished product ideas in the hope of selling to your company, and you see an opportunity to incorporate them into your product. For example, giant machine tool company K invited me in to discuss incorporating my Cartesian robots into their systems. After extensive disclosures and quotations they announced to me that they would make Cartesian robots themselves and no longer needed me. I sued for unfair business practice, and they settled the suit with an agreement to maintain my confidential disclosures confidential for a period of time. That was "all the justice I could afford" in a legal, money-bleeding contest with a very big company.

Is it ethical to invite in a vendor's salesperson just to learn something when you know ahead of time that you will definitely not buy from that vendor? International Business Machines (IBM), my favorite customer, forbids the practice. It teaches and actually enforces a strict code of ethics on all its employees who deal with outsiders.

6. How honest should you be with vendors when you are encouraging them to spend effort in trying to sell to you? When I was employed in the space business, I responded to a government request for proposal (RFP). The government agency liked my proposal so much that it canceled the RFP and issued a new one with copies of my drawings; a different vendor won the contract on price. I regret that this story is common.

7. When a patent expires, it is absolutely legal to copy the product in detail and sell it. How do you feel about participating in such a process?

8. How do you feel about circumventing the security system in your own organization, not to steal secrets or property but either to see if you can do it as a challenge or for your own convenience?

9. How do you feel about circumventing the procedural rules of your organization in order to make faster progress with your work?

When I was designing Roamer, the proposed unmanned moon surface vehicle (Fig. 2.1), a very capable engineer, Nellis Hill, worked with me and did such a remarkable job of getting the model built by our large organization that this experience was one of the reasons I later hired Nellis into my first company. One day, while reminiscing, Nellis said, "Larry, remember when we were building Roamer and I used to give you forms to sign every few days? You had no authorization to sign those forms, but I needed a signature to get the work done, so I tried yours and it worked."

10. Is it proper to moonlight for another company (presumably not competitive with yours)? Is fatigue a consideration?

11. Should you blow the whistle on a person or an organization, including your own, when you become aware of serious wrongdoing? Should such whistle-blowing be equated with informing and talebearing? In some government contractors' plants the Department

of Defense (DOD) requires a hot line for reporting security, fraud, waste, and abuse violations.

As a consultant I designed a fast material-handling system for client L, who needed it to make his ultrafast fabricating machine work for customer M, who would use the fast machine to replace many slow machines and their operators. I was familiar with M's plant because I had visited it when I helped sell the contract.

So far, all three parties, L, M, and myself, were on the side of the angels. Then, one night, I realized that the fast handling system would permit a radical speedup of the old machines, which had really been slowed down because of slow material handling. I told this to my client, the president of L, who told me to forget it; they were in the fast-machine business, not the materials-handling business. I felt that I was now an accessory to fraud and blew the whistle. Of course, my client threw me out. Did I do right? Years later someone suggested a third alternative: I could have quit and walked away in silence. At the time, I had no time to meditate on the problem. The purpose of this chapter is to give you time in advance of the problem.

12. You will be called upon to give personal referrals about people you know and may have worked with. When your opinion includes an unfavorable portion, do you say it like it is, do you distort or lie, do you evade the unfavorable portions (which is a silent lie), or do you refuse to give referrals? Many personnel departments will say nothing except to confirm employment dates. What is your ethical responsibility to the inquirer and to the person inquired about? Suppose the inquiry is from the Federal Bureau of Investigation (FBI) as part of a security check? Suppose it is from the police as part of a criminal investigation?

13. Bootlegging, i.e., working on a project which has not been authorized by your managers and charging the work to an account for which you are not working, is a wonderful subject for engineers. The practice goes on in many sizes and types.

You can bootleg a project for your organization when you have been denied authorization and you feel that it is important that the project be done or when you want to get it started before the authorization paperwork has been completed.

You can bootleg a project for an outsider either for a worthy cause or to sell your time and the organization's resources and pocket the money. [I bootlegged an entire heart-lung machine (Fig. 2.3) for some doctors with the connivance of the company president, the model shop, the purchasing department, and everyone but the company's customer, who paid for it.]

You can bootleg a project for yourself. This can be anything from making a personal phone call on company time to using a company drill press to drill a hole in a part you own, to building a large personal project with your own company time, assistants' company time, and company materials and equipment.

You can bootleg a small amount of company time to think out, discuss, and write a proposal to the company that it undertake a project or change a practice. In some organizations there is an authorized account to charge for such time, in which case it is not bootlegging.

The magnitude of a bootleg project can vary from a few minutes of your time (that personal phone call) to the development of the Sidewinder missile (which was bootlegged by a brilliant Navy laboratory director). Clearly, to do zero bootlegging is quixotic, so your ethical question is where do you draw the line? (There is also the question of how brave you are, a question for both good guys and bad guys.)

The biggest and best and most successful bootleg job I know of was the Sidewinder missile. Bill McLean was technical director of the Naval Ordnance Test Station in China Lake, California, when he invented the Sidewinder. However, air-to-air missiles were Air Force turf, and such work was strictly against Navy rules. McLean had the legal right to spend a large budget as he saw fit, subject to Navy rules. He was a tremendously capable man with a superb record of achievement and a corresponding self-confidence. He did the development in secret defiance of the rules. It was a tremendous technical success. There were screams of protest afterward, but his success prevailed.

There is an authorized form of bootlegging in military contracts called independent research and development (IR&D). The contractor is authorized to work on anything he or she wishes, relevant to the general field of the contract, up to a percentage of the contract price. The intent is to encourage preliminary work which will lead to proposals for new contracts.

I was at a party in Washington in which there was a heated argument among a group of managers at the National Bureau of Standards (NBS). Each insisted that he was "the biggest crook at this party." I knew them. None would take home a paper clip, but each had diverted large amounts of budget from what he considered worse uses to what he considered better uses.

14. Credit for an idea is a sensitive subject for design engineers. The problem is not whether or not you take credit for someone else's idea (that is a simple moral policy decision) but how you handle credit for ideas to which more than one person contributed. A contribution may be the statement of a problem, the asking of a question when an early version of an idea is presented (leading you to improve the idea), a criticism of an early version of the idea (leading you to improve the

idea), a minor suggestion, a major suggestion, joint development of the idea, and, down the other side, a case in which the other person is the principal originator and you are the contributor.

A special case of idea credit sharing occurs when the idea is to be patented. Patent laws make no allowance for good-fellowship; if the inventor statement is untrue, the patent is invalid. Furthermore, the inventors must sign, under oath, that they each believe that they are the true and legal inventors. What do you do when someone else, either in good faith and bad memory or to muscle in on your credit, claims to be a coinventor? (I once worked for a company in which the engineering vice president "coinvented" every invention for which a patent was applied.)

15. What is the ethical response to your supervisor's explicit directions and implicit policies when you disagree with them? Do you silently obey with all your ability and enthusiasm, do you first argue and then obey, do you obey in a halfhearted way, do you evade and procrastinate, do you refuse to obey, or does your action depend on how strongly you disagree? What is the basis of your disagreement, the company's interest or your own, if you are able to make the distinction?

16. I have observed a hierarchy of loyalties in organizations. A typical one is: first to oneself, then to one's friends, then to one's immediate group, then to the next echelon up, and so on with the stockholders or the country or the board of regents at the end of the line. How should you handle conflicts of loyalty on this ladder? (I once had a manufacturing manager who fought every cost reduction I tried to introduce because it threatened the job security of his friends whom he had hired into the company.)

17. What ethical feelings do you have about joining a labor union? Is this an ethics question?

18. What ethical consideration do you have about designing military products? Does it vary with the type of product? How about medical devices, uniforms, trucks, conventional weapons, nuclear weapons, Star Wars, chemical weapons, poison gas, bacterial weapons?

19. What are the ethics about using your employer's confidential data (design ideas, drawings, data, customer data, etc.) in starting a new company? Suppose the new company does not compete with the old one? Suppose it does? Suppose you don't start a new company but use the data to get a job with a competitor?

20. Many people rationalize their self-interest until they believe that "what is good for me is good for the organization." Do you have an ethical requirement to follow Socrates' injunction to "know yourself"?

21. Should you adjust the numbers on your travel expense reports

to compensate for improper limits by your company on hotel, meal, and other allowances?

Finally, if you like the subject, I suggest you get acquainted with the game of Scruples.

REFERENCES

1. Martin, M., and R. Schinzinger: *Ethics in Engineering,* 1st ed., McGraw-Hill, New York, 1983.

Your Career

Planning

You can take the first job you can find, then drift from job to job as your supervisors direct and as opportunities arise. To some degree you will have to do these things whether you like it or not. On the other hand, you can at least influence your career by deciding in advance your preferences among alternate career branch points, and you can help it in a large number of specific ways. Regardless of your career path, there are some facts, ideas, and specific recommendations that I think will increase your success.

Understanding Yourself

One of the best things you can do for your career is to learn to under-stand your *real* reasons for your thoughts, arguments, and actions. If you are human, they are often reasons of feeling and emotion rather than reasons of logic. In dealing with others, one of the most effective things you can do is to understand the other's *real* reasons, which may be very different from his or her expressed arguments.

Branch Points

Generalist-specialist

There is a branch point for you between generalist and specialist. (A generalist is one who knows less and less about more and more until he or she knows nothing about everything; a specialist is one who knows more and more about less and less until he or she knows everything about nothing.)

Generalists suffer from the anxiety that they do not have the depth of knowledge that specialists have on any subject they deal with. On the other hand, they can bring to bear on a design problem a diversity of knowledge which the specialists know nothing about, and they may have an insight into the realities of the problem (conditions of service, general specifications, manufacturing, available materials and components, forms of power and energy, etc.) which specialists may lack. They also have the economic security that they will not become professionally obsolescent overnight by the announcement of a new invention or the unexpected termination of a program's funding or some of the other shocks described below. The best academic program for future generalists is engineering science rather than any branch of engineering.

Specialists are on the cutting edge of their technology and may well advance that edge. They need fear no one's pointing out gaps in their knowledge and skill. They are sought after because of their definable expertise. If their field is in demand, they may command high pay. They are usually proficient in mathematics and use mathematics and computers extensively in their work. They have up-to-the-minute knowledge of what their competitors are doing. But all this is only in their specialty because they lack the time to acquire the knowledge and experience of working in other fields.

These descriptions are of the end points of a spectrum. There are real people with all possible combinations of breadth and depth. The point of the discussion for you is that you should establish a career policy to suit your aptitudes and tastes on this and the other issues described below and then bias your decisions, your requests to management, and your continued technical study and reading to move yourself in accordance with that policy.

Let me warn you of another form of specialization. In a large organization doing complex projects such as airplanes, it is efficient to assign individuals to narrow fields so that the overall collection of specialists has a deep knowledge of all relevant fields. Unfortunately these fields can be very narrow indeed so that engineers can find themselves specifying cables and connectors year after year. I once interviewed such a man. After basic technical questioning he left my office in tears, saying that he now realized he was no longer an engineer.

On the other hand, a friend of mine was assigned the task of specifying transducers for the Atlas program. He does no design, but he knows all about available transducers, has written a reference book on the subject, which he periodically brings up to date, and he now works in one of the space program laboratories. I sometimes consult him and have referred my clients to him. He's completely happy with his career in transducer specifications.

High tech–low tech

There is a branch point among high, medium, and low technology. Your choice depends on your own judgment of your tastes and aptitudes. The world needs all three. Apply to companies in whose technical level you will be comfortable and successful.

Technical work–management

The next career branch point is the choice between technical work and management. Almost all training in engineering schools is technical. M.B.A. schools teach management. Both schools leave the other branch's knowledge to be picked up at work, at great cost to all concerned except the schools.

As an engineer you deal with a world which is primarily physical. Much of this book deals with the human world you must also face. Managers deal primarily with people, politics, and administration and secondarily with making or approving decisions requiring technical judgment. Managers have more power, more prestige, and better offices and secretarial help, make more money, get to talk to senior managers, feel important, and are on a promotion ladder with a higher top. They are less likely to be trapped in a narrow specialty without wanting to be a specialist. (A class of manager is the engineer-entrepreneur who is lord of all he or she surveys but who may not survey very much.) All these benefits tend to draw engineers into management with a mixed bag of successes and failures.

Some large companies have reduced the transfer temptation by setting up an engineering ladder parallel to the management ladder with titles, pay, and badges of rank to match rung for rung. I was a "design specialist" in Convair with good pay and the red badge of management and was quite content with the prospect of a future promotion to staff specialist. (After working successfully for several years, I happened to see the personnel department's job description of design specialist. Einstein and Edison together couldn't have handled the job.)

The aptitudes for management are as different from the aptitudes for engineering as are the contents of the two fields of work. For engineering management you need both sets of aptitudes. Do you have them? (I have not met you, so I don't know if you do. But it's extremely important for you to decide if you do.)

The Peter principle refers to a person being promoted with changes of duty until "he reaches his level of incompetence." In your own interest be careful that you don't seek or accept promotions to positions for which you do not really think yourself qualified.

I must tell you a Peter principle horror story. Over the years in my small companies I have promoted two excellent engineers who were working for me already, and hired two other engineers who seemed good on interview, to be chief engineers of small engineering departments. All four were disasters. Each made incredibly costly decisions and had to be removed. Perhaps if they had been promoted in small increments in a large company and had been closely supervised and trained by senior managers who were themselves trained to do so, they might have been successful.

The other face of this coin is that great and prolonged stress can bring out and develop capabilities in persons that they would have denied having. When I started my computer terminal company, I did so with the announced premise that I absolutely could not do marketing, was afraid of it, and would start the company only if I had a proven marketer on the team. I was assured by my cofounder that F was such a man. After unbelievably incompetent performances by F and by two successors, there I was on my own with a worthless marketing organization and little else. It was sink or swim, so I swam. I have been my own marketing manager ever since, with enough success that I could retire and write this book.

Big company–small company

The next branch point to discuss is big company versus small company. The advantages of working for a big company are:

- The magnitude and complexity of the projects are larger.
- The facilities are greater: computer, laboratory, shop, and office equipment and their operators.
- The corps of experts whom you can consult and from whom you can learn is greater.
- Career guidance by capable supervisors may occur.
- There are tall ladders of promotion and pay which you may be able to scale.
- There is public prestige in your association with a famous company name.
- The perks and benefits (level of medical insurance, payment for jury duty, stock plan, pension, etc.) may be greater.

The disadvantages of working for a big company are:

- Greater conformance to rigid rules by accounting, personnel, purchasing, and other departments is required. There is much more paperwork.

- There is greater compression into narrow specialties and small fractions of an overall program.

- Slower results may be caused by having to follow fixed procedures and delays in work that you delegate to other departments.

- It is much harder to exercise initiative except in your assigned narrow area because of the organizational specialization and the long list of approvals required to authorize something new.

The advantages of working for a small company include the reverse of the disadvantages of working for a big company. In addition:

- You get to know more of what is going on besides your own assignment.

- You are known to senior management, and you know a large fraction of all the employees.

- Your assigned responsibilities are broader.

- You can grow in scope and responsibility much faster. Reviews may come when they are appropriate, not on the calendar.

The disadvantages of working for a small company are largely the reverse of the list of advantages for the big company. However, career guidance and teaching depend on how lucky you are in your particular supervisors and associates.
Some benefits and penalties are independent of company size.

- Job security depends on the individual company and its business success, not on the company's size. Success and failure come to all sizes. Very large corporations rarely go bankrupt, but when a program or a product line shrinks, its staff gets laid off. It happened to me.

- Level of technology depends on the individual company. Many small companies are started and succeed because they are the leaders in the development of an advanced technology. Many large companies make technically elementary products, with slow technical progress, but make and sell them in huge quantities.

- Pay scales for the same job vary little with company size but may vary significantly with industry. Professional society and trade journals periodically publish this information and related data.

I have worked for small, medium, and large companies. I prefer the environment of small companies, but that is for my own tastes and aptitudes. People have been successful and unsuccessful in all sizes of companies. I can make no general recommendations.

Type of organization

The next branch point to consider is the type of organization you prefer.

Universities do research, and their laboratories require the design of unconventional devices of great diversity, usually made in quantities of one. Fringe benefits may include lower cost and greater access to higher education and association with research faculty, who may teach and encourage you as well as being highly educated and interesting people. Some university laboratories are very large and are, in effect, government laboratories by contract.

For example, my wife joined a research group at Brooklyn Polytechnic Institute as a technician with a B.A. in mathematics and some elementary training in radio. The group became a leading microwave manufacturer, she was encouraged to do graduate study, and when we were married, she had a master's degree in electrical engineering and was an executive of the company.

There are many government R&D laboratories as well as government engineering operations other than R&D. Most are federal, but some are state or even municipal. Facilities, diversity of work, and magnitude of projects vary from very small to greater than those of the largest companies. Lifestyles vary from intense high-level technical work by very able, energetic, and committed people, through jobs with more administration than engineering, to the infamous bottom layer of loafers whom it is too difficult to fire under Civil Service rules. In some ways government jobs are more rigidly bureaucratized than jobs in the largest companies, yet the largest bootleg project I know of, the Sidewinder missile project, was done in a government laboratory.

There are research and consulting companies of many sizes that sell and perform contracts, some of which require design engineering of many kinds. Such companies may provide the greatest possible diversity of projects. I have worked for two and started one. Unfortunately, job security may be poor because if the company runs low on contracts, you may be laid off.

The last and largest category is manufacturing industry, which we have already discussed.

Large projects–small projects

Another branch point is whether you like to work on large projects which require a long time for completion but are very impressive when done or on small projects which show commercial results in a short time. (An airplane is a large project; a toggle switch is a small project.)

It may or may not be important to you where you live. Family ties, climate, and nearness to big cities or to recreational areas are all matters of great or small importance to different people. If it matters, apply only to companies whose locations qualify for your taste.

Finding Your Job

Unless times are hard and you must apply for every job you might get, you can decide which size is best for you and select your ad responses and unsolicited applications to suit.

You are not limited to answering ads or employment agency referrals. If you know the kind of company you want to work for, you can get company names from *Thomas Register of American Manufacturers*[2] in your public library and from the chambers of commerce in the cities in which you are interested. I found my first job through *Thomas Register.* It is the Yellow Pages of American industry. There are other industrial directories also.

You can learn a great deal about a prospective employer by reading the financial publications about it: annual report, analysts' reports, *Moody's, Standard & Poor's.* (See Chap. 12, "Your Company.")

Your résumé

Permit me to give you some advice about your résumé. I have read hundreds, looking for good engineers, and I know what makes a good impression and what is a time-wasting bore. Remember that your résumé is your sales brochure and that it must persuade the reader to grant you an interview.

Do not include your height and weight, photograph, hobbies, family data, or social background. Your reader wants an engineer, not an in-law.

Do put in every single fact that is relevant to your professional competence:

- College and professional education and degrees and academic standing, if high. Include nondegree courses.

- Academic and professional awards (e.g., cum laude, Tau Beta Pi).

- Licenses (e.g., P.E., patent agent, radio operator, pilot).

- Employment history, last job first (dates, employer, tabular list of all projects or jobs performed). Do not include salary or reason for leaving. These will be discussed and negotiated at your interview and will not influence whether or not you get the interview.

- Tabular list of all patents.
- Tabular list of all technical papers and articles.
- Tabular list of other professional qualifications not included above.

If you are a new graduate, you don't have anything but your academic record. Recite your courses and credits. A photocopy of your transcript says it all and shows your candor. If you were not a straight-A student, there are still jobs for you and grades are only part of what employers look for. Recite extracurricular activities of professional relevance. (Building solar-powered vehicle, yes. Ski team, no.) If you worked your way through, say so; the fact is a predictor of a responsible, hardworking employee. Give a very brief statement of the general professional field you want to enter. If you plan advanced-degree work, say so.

Don't fear length. The more pages of facts, the better. (My résumé now runs 16 pages after 47 years of accretion.) There are places where the one-page condensed version is useful (e.g., mine in this book), but if you are going for a job, the more meat the better. Remember, I said "meat." There must be no padding whatsoever. The very worst thought you can put into your reader's mind is "phony."

I urge you to read Ref. 6 in Chap. 4, in which a very successful advertising man tells you the rules for writing "potent copy." In brief he says, "Write facts, lots of facts, nothing but facts." You may be offended by the idea, but your résumé is really your *advertising brochure*.

Interviews

When you are interviewed, ask questions as well as answer them. Have a list of questions ready. You will gain respect for doing so. Be neither arrogant nor humble. Genuine self-respect is impressive.

Give free samples if you can. I have had many interviews in which I sold consulting engineering. Such interviews are almost the same as job interviews. I always try to learn about current problems and do as much engineering on the spot as I can. My interviewer can then see the real thing, not just hear promises.

Decide on your willingness to "dress for success" (Ref. 4 in Chap. 4). If you feel that you should dress to suit yourself and it's none of anybody else's business, choose geographical areas and companies whose culture accepts that. I must tell you, however, that engineering is a rather square profession, and conforming to an unwritten dress code can help you.

Shocks You Can Get

In any job in any organization you may get unexpected, unearned, and violent shocks. I advise you to make contingency plans so that you will be ready if one occurs. Among these shocks are:

Layoff

Layoffs occur when a contract is concluded or canceled, manufacturing business declines, a plant is closed and its work transferred, or management decides that a program should be canceled for any number of reasons of its own: policy change, merger, reorganization, etc. The aerospace industry is infamous for giant swings in hiring and firing. People are laid off en masse regardless of merit. A few may be selected for transfer to other programs because of their recognized merit. (One of your plans for success might be to work very hard to develop a lot of recognized merit.)

I was the first person hired by a middle-sized company in a planned expansion. After 5 years, during which I was quite successful, the expansion was canceled because the accounting department priced us out of competitiveness, and I was laid off with everyone else.

There are small-scale layoffs too, to clear deadwood, when business slows down. With a little laziness, negligence, and personal unpleasantness one can qualify oneself as deadwood.

Technological obsolescence

Technological obsolescence is an ever-present threat to the narrow technical expert. It takes the form either of failure to keep up with advances in your field or of obsolescence of your art. The need for the developers of vacuum tubes and of mechanical calculators decreased quite suddenly with the advent of the transistor and the electronic computer.

For example, I once invented the Decimal Keeper slide rule (Fig. 6.1). I divided the length of a pair of scales, C*, D*, into 20 decimal cycles and labeled them with exponents -10, -9,..., -1, 0, $+1$,..., $+9$, $+10$. The user no longer figured the decimal point by approximation with pencil and paper but simply ran the problem a second time on C*, D*, entering each number with its decimal point and

Conventional side

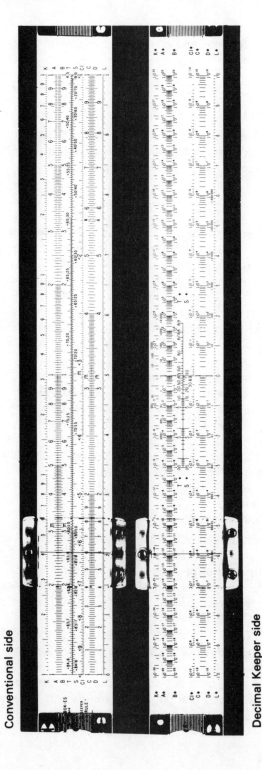

Decimal Keeper side

Figure 6.1 Decimal Keeper slide rule. (*Courtesy of CPG International Corporation.*)

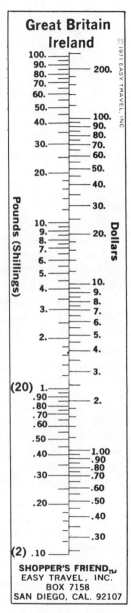

Figure 6.2 Shopper's Friend currency converter.

reading out the answer with its decimal point. The results were good to between one and two significant figures. For approximate work one pass through the Decimal Keeper scales was sufficient. The advantages were fewer errors and eliminating distraction from the main problem caused by thinking about the decimal-point calculation.

The Patent Office granted a patent, I licensed one of the major slide rule manufacturers, and it manufactured and sold the device. I had a mail-order business at the time, became a distributor for the slide rule company, and started to make significant sales.

One day I saw in a store a new little box with digit buttons similar to a telephone's, $+$, $-$, $\times$, and $\div$, buttons, and a row of luminous digit displays. I sold off the inventory and dropped the whole damn thing. Technological obsolescence!

Your job can be lost from government party switches. The other party, coming into power, moves budget items up or down. International agreements change defense budgets.

For example, at the same period as my slide rule adventure I invented the Shopper's Friend for tourists, had it manufactured, and sold it mail-order and through luggage shops and travel agents (Fig. 6.2). It provided, for each country, a nomogram matching that country's money with dollars in the range $0.10 to $100.00. A tourist could then tell at a glance the approximate dollar price equivalent to the local money price with sufficient accuracy to make a buy decision. This, too, had started to sell well when, one Monday morning, the newspapers announced that the central bankers of the world had met over the weekend and decided to let the money ratios float! Another instant obsolescence. (When I tell the story, I usually start by saying that I was the victim of a conspiracy of international bankers.)

New boss

A common shock is the new boss. You have been happy and productive with your manager. Then the manager disappears by promotion or new job or sickness or retirement, and suddenly there is a replacement and you don't get along. In the aerospace company I was reassigned from a wonderful boss to a thoroughgoing SOB. I went to the next senior manager and asked for permission to apply to the sister division. Knowing that this was always frowned upon, I had decided, and said, and meant, that if permission was refused I would quit. I got permission, got two job offers in the other division, and spent the next year and a half happily inventing and writing proposals and technical papers in the space business. (I helped propose and sell the company's first satellite contract.)

Happy shocks

One last word about shocks: there are happy ones too. You may get a surprise promotion or assignment because there is a sudden opening and the manager has had an eye on you. On the chance that this might happen, you could do worse than to prepare yourself and to look good to that manager. I've had it happen to me.

What Should You Do about It?

Network

Actively build and maintain a network of professional and business friends and contacts. Join your technical society and be active in it. Attend technical meetings and make friends. Give papers. Make the branch-point decisions described above. Regularly study trade journals and newspaper help-wanted ads and company catalogs and have a shopping list ready, just in case.

Build your résumé. In making study, work, society membership, and other decisions, think whether the results will appear favorably in your résumé.

Study persuasion

Really study the Golden Art of persuasion (Chap. 4). It will help your career every single day.

Save money

You can shop longer for a better job when you don't desperately need a job. Furthermore, your self-confidence is hurt by financial anxiety and helped by the feeling of financial security. In turn, your job security is helped by your appearance of self-confidence and hurt by your appearance of anxiety.

The greatest case I know of the benefit of independent economic security was engineer G. G grew slowly as a boy, became a jockey, and learned horse racing. When he outgrew the body-size limits for jockeys, he went to school and became an engineer; G happened to be very bright. G used to take groups of other engineers to the racetrack, where they always won because G could predict which horses would be pulled back and which would be turned loose. G worked for an engineer's salary when he could get rich at the track because he had two teenage sons and his wife insisted that they have a respectable father. No manager ever mistreated G, and his demeanor showed that he knew what was what. Also he was bright and energetic and did very good work because he wanted to.

It will not come as a total surprise to you that it does not hurt your job security if you work hard, do advanced study, and are a cooperative and likable person.

Improve your qualifications

Improve your qualifications for other jobs and, at the same time, for promotion in your present job:

- Study for an advanced degree, at night, if you think you can handle the work both academically and within family obligations. If you have management or entrepreneurial ambitions, get an M.B.A. Consider whether you can finance full-time day study. If you can, it is easier, does a better job, and gets you there sooner.

- Get your P.E. license.

- Take extension courses and professional society courses. Your company will probably pay the tuition, but this is a good investment even if you pay.

- Read a lot: technical books, other nonfiction, fiction. At the very least, your supervisors and prospective employers will recognize you as an educated person and respect you for it.

For Women Only

I'd like to add a few words specially for women engineers. You have a competitive advantage over men when you first graduate because of affirmative action. However, the tendency is to hire women but not to promote them very fast, if at all. Try to learn your prospective employer's track records in this matter.

Your prospective employer would love to know your plans regarding marriage and children, but it is absolutely illegal to ask you. It is not illegal for you to tell voluntarily, however. If you plan to suspend your career very soon to raise a large family without nursery schools, you need not say anything about it. But if you plan to take short maternity leaves and get right back to work, it won't hurt you to volunteer the fact. When my daughter had one boy and one girl in nursery school and had decided not to have another child, she actually took my advice and put those facts in her résumé, and they helped.

If you want to participate in the equality struggle, join any women's organizations you choose, including the Society of Women Engineers, but do not help create or join a woman's technical society, or you will be perpetuating your status as a second-class citizen. Join a regular technical society as a member, not as a woman.

I do not have any knowledge or opinions about how being a woman should bias your branch-point decisions. I wish I could help more. Reference 1 may be of considerable help to you in many ways.

Sales Engineering

I would like to put in a plug for sales engineering as a career. I have done sales engineering a great deal and have found it highly rewarding.

A sales engineer is not a salesperson. You do not knock on doors, cultivate purchasing agents and secretaries, buy lunches, and ultimately bring back the order. The sales engineer is technical staff to the salesperson, who must do all those things. You are introduced by the salesperson to technically qualified opposite numbers in the customer's organization. Your job is to teach the product, apply the product to solve the customer's problems, which can take high-level engineering, advise the customer how to modify a problem to be able to use the product to advantage (this is called "consultative selling"), give the customer whatever other consulting assistance you can (thereby enhancing the prestige of your own organization), write a technical report of the visit, and write a technical proposal if appropriate. The salesperson takes over from there.

The work provides wonderful opportunities to propose improvements in your own products, to devise new products, and to see the real world of people, products, and factories.

REFERENCES

1. La Rouche, J., and R. Ryan: *Strategies for Women at Work,* Unwin Paperbacks, London and Boston; also Avon, 1st ed., New York, 1984.
2. *Thomas Register of American Manufacturers,* Thomas Publishing Company, New York. Revised annually.

Efficient Use
of Your Time

How to Play Office

Here is a list of techniques to help you avoid getting down to productive work, the effort it requires, and the stresses it causes. These are the rules for "playing office":

- Morning start-up coffee, preferably with sweet rolls and conversation. Multiple trips to the coffee source during the day, with conversations more or less related to the work at hand.
- Busy work: Rearrange your papers. Rearrange your office tools. Get up and get a fresh pad of paper or anything.
- Read something unnecessary.
- Try something new on your computer terminal.
- Discuss something unrelated to the work at hand, such as how the company should be run, or the country.
- Write an unnecessary memo and discuss its typing and distribution, at length, with the secretary.
- Make a photocopy of something.
- Hold a meeting about anything. Discuss everything in the meeting.

- Work on the easiest part of the job first, and save the serious problems for last.

- Make arrangements for lunch.

- Be serious and careful with administrative paper: time cards, expense reports, requisitions, work orders, etc. Do not delegate filling out forms to secretaries, drafters, technicians, or other assistants; this work is too important to be done by anyone other than yourself. If your company is suitably organized, you can spend most of your time in this activity, fritter away very little time producing actual engineering, and qualify yourself as good management material.

- Remember the basic rule of work time: to accomplish 1 hour of solid work takes between 90 minutes and 3 years, depending on what the worker considers more important.

How to Produce Engineering

If you really want to produce a lot of engineering as your path to success, I can make a few suggestions:

- Spend a small amount of time in planning your time. Use **PERT** charts or other scheduling charts, but remember that they are to help you work; they are not the work you do.

- Decide priorities and follow them.

- Devise rules and habits to avoid the avoidance techniques listed above, and practice quiet self-discipline to make these rules and habits work.

- Plan on paper and review your own performance.

- Make checklists and check off the items. (Some people are so compulsive about checklists that they will enter items already done so that they can have the pleasure of immediately checking them off. I am guilty of this myself.)

- Get one of the many commercial record forms and planning calendars you can buy to help you, but remember that they help only if you use them:

Month-At-A-Glance

Week-At-A-Glance

Day-At-A-Glance

Printed by Sheaffer Eaton Division, Textron, Inc.; sold in stationery stores. (The names are trademarks of Textron, Inc.)

Day-Timers

Day-Timers, Inc.
Allentown, Pa. 18001
Telephone: 1-800-233-6864

Day-Timers also makes a financial records book called *Myfar* which we use for family finances. (The names are trademarks of Day-Timers, Inc.)

Franklin Day Planner

The Franklin Institute, Inc.
P.O. Box 25127
Salt Lake City, Utah 84125-0127
Telephone: 1-800-654-1776

Filofax notebook and forms, sold in stationery stores. (The name is a trademark of Filofax Ltd.)

My own practice for many years has been *Month-At-A-Glance* for scheduled events and list after list for things to do.

- Delegate. This is easy to say but hard to do successfully. The keys are your judgment of the delegatee's competence and diligence and your ability to persuade him or her to perform (see Chap. 4, "Persuasion: The Golden Art"). If the delegatee is successful, you have delegated well, by definition. If the delegatee is unsuccessful, you have delegated improperly, by definition. Lots of luck!

- Do the unpleasant tasks first. Then you won't procrastinate, to your disadvantage, and besides you will feel virtuous.

- Be neat. I believe you will be more successful by being neat because work will not be lost or overlooked, feelings of confusion will be reduced, and concentration will be made easier. Visitors will be given a feeling of confidence. In a laboratory and in a shop, safety to people and to equipment is greater with good housekeeping.

There are companies which carry the neatness rule to extremes: no more than two pieces of paper on a desk at any time. On the other hand, a typical magazine editor's desk carries a mountain of papers, and the editor will look you in the eye and declare that he or she knows where every paper is in the mountain.

My own practice is to maintain a neat office, encourage office workers who work for me to do the same, and absolutely insist on neat housekeeping in laboratory and shop.

As I said, I believe office neatness will help you, but there are many who disagree, so I do not make office neatness a positive recommendation.

Secretaries

Try very hard to get and keep a good secretary. A secretary who takes filing and neatness seriously, is courteous and tactful to others than yourself, who is bright and energetic, and who wants to be helpful is an absolute blessing. Such a secretary is an assistant, not a clerk.

As a new hire right out of school you will wait hat in hand to get a secretary to type a letter. When you get to be a big wheel, you will have a secretary all your own. In between you get some share of secretarial attention. I have been fortunate in most of the secretaries I have had, and at this point I want to thank them again with absolute sincerity and gratitude.

The other face of that coin is that you should treat secretaries as VIPs and earn their goodwill. A secretary who does not like you is a bad enemy, especially one who is the secretary to your boss or someone else important to you. Try to make up. If you cannot, let the secretary's boss know of the hostility so that he or she can discount the things the secretary will say.

Allocating Your Time

In planning your time remember that thinking is working. Do not feel guilty if you are physically immobile and are not spreading ink and graphite over paper or electrons over a fluorescent screen. A knowing supervisor of a good engineer is delighted to see the engineer with feet on the desk and a scowl on the face. On the other hand, if you work for the Bull-of-the-Woods, learn to think while bent over a desk with a pencil in your unmoving hand.

Remember that you are using your time competitively with someone else using his or hers, whether inside your own company or inside a competitive company. Whichever competitor uses time more efficiently will get a payoff. Alexander Graham Bell beat Elisha Gray to the Patent Office by only a few hours and was awarded the basic telephone patent because he did.

You will always have competing demands on your time from your job, your continuing education, your family and social life, your civic affairs, your exercise, and your rest and recreation. The ancient Greek concept of the golden mean suggests that you allocate a "reasonable" portion of your time to each. On the other hand, some people are driven to put business and professional success ahead of everything else. I won't tell you which to do, but I will advise you to take some time to think out and decide your policy on time allocation. Don't just let it happen.

When I first started consulting, I had the opportunity to put in all the hours I wanted to for a former employer at what then seemed a princely hourly rate. My first impulse was to work day and night and get rich quick. I soon put that program aside for what seemed a balanced mix of work and play, but from then on I have never *wasted* time in what I call "piddling around," neither working nor playing.

In this book I recommend very many things that you should do to increase and speed up your success. You have neither the time nor the energy nor the dedication nor the ability to do everything you should to the fullest possible extent. I certainly have not.

One option is simply to try to do reasonably well in all directions as corresponding needs arise. Then you accept the imperfect results as they occur with some combination of frustration and apathy, depending on your personality.

Another option, which I recommend, I call "selective neglect." You don't decide only what you will do; you consciously decide what you will not do, i.e., what you will neglect. You predict the gains from what you choose to do and the losses from what you choose to neglect. If you judge your predictions well, you will maximize the efficiency with which you use your time, i.e., the amount of success per unit of time spent.

Let me end this chapter on a lighter note. A British professor in the University of Singapore (yes, Singapore) named C. Northcote Parkinson (yes, C. Northcote Parkinson) has devised a series of "Parkinson's laws" dealing with the efficient use of human time, both by individuals and by organizations (mostly the British Admiralty). He has written them up, with examples, in several slender volumes listed in the references of Chap. 17. I urge you to read them for both entertainment and insight.

8

Coexisting with Accountants

Most design engineers react to accountants as a mongoose reacts to a snake. "Bean counter" is the mildest epithet used. Yet it helps to understand where accountants fit into the scheme of things so that you can deal with them effectively and even benefit from their activities.

What Do They Do?

Managers hire accountants as *reporters* of our "Pervasive Parameter: Money," Chap. 20. Accountants divide the money they watch into categories called "accounts" and *report* where money has been spent and what remains. The list of categories is called the "chart of accounts," and much good and much harm can be produced by a properly designed chart of accounts. A misleading classification for R&D once cost me, personally, several million dollars in establishing the sales value of one of my companies, and I would like to do some very illegal things to the accountant and bookkeeper responsible.

To some degree accountants follow tradition, which they could change if they wished, and to some degree they are forced by law to obey a rule-setting organization called the Accounting Practices Board (APB). The federal Securities and Exchange Commission (SEC) and the various state corporation commissioners also do rule making. The accountants' topmost reports to your management, and to the public if your company is a public company, are called the "profit and loss statement" and the "balance sheet." The categories in these documents are rigidly defined by the rule makers.

One of the sins of the current rules is that profit and loss are reported and announced quarterly with little discrimination between real losses and expenditures made for long-range profits. Guess how this practice motivates managers who are judged on these quarterly reports.

The second category of report which the accountants must produce, by law, is the computation of taxes. This report, to a large degree, is an interpretation of the profit and loss statement. Such interpretations permit considerable leeway to the accountants and offer an invitation for what that fraternity refer to among themselves as "creative accounting."

Accountants are reporters like journalists and, like journalists, commonly recite the raw facts with reasonable correctness but without understanding. It is this lack of understanding and poor interpretation of the facts which is a principal cause of conflict between engineer and accountant.

Another major source of friction, neglect and inaccuracy on our part in presenting the accountants with the correct numbers, is pretty hard to defend. We engineers should take such reporting as a necessary evil and do our duty.

Throwing Their Weight Around

Like everyone else, accountants like power. This shows, in an entertaining and harmless way, in their use of the word "controls." As a sometime builder of servomechanisms, I was bothered by this usage for years, since accountants really can't control much no matter how they try. The explanation is that "control" is accounting language for "report." Reporting money facts is precisely what accountants are supposed to do.

Another way that accountants, successfully, reach for power is in influencing management decisions with concepts of "burden" and "overhead" as real and intrinsic percentages of productive costs and persuading managers to act with these concepts in mind regardless of the reality of their relationship to the real world. I was once part of a layoff when this process killed a company division dead. Our grandchildren will continue the struggle.

Yet another way in which accountants exercise power is in the design of business forms and procedures, such as expense reports, requisitions, etc. I think accountants really believe that, if you divide your expenses into travel, car rental, airfare (all different), hotel, meals, etc., tabulate by the day, add each day's expenses vertically, add the categories horizontally, add the category subtotals vertically, add the day subtotals horizontally, and come out with the same number in the lower right-hand corner, it proves you didn't cheat.

The first thing I did in my first company was to institute the engineer's expense report form. It is just a plain envelope. You write your expenses on the front, in any order, and put your receipts inside. You toss the envelope on the bookkeeper's desk, without even adding it up, and go back to your real work. The bookkeeper gets paid less than you do to add better and to sort receipts into categories without hating the task. You have engineering to do.

One of the worst things these people do is to persuade managers to institute a conspicuous "overhead" cost reduction at the invisible expense of increasing the cost of the real work. I have seen them "reduce" long-distance telephone cost by making calling so difficult that engineers have done without the telephone to avoid the bother. (We were advised to write letters instead.) I have seen them eliminate a messenger service by minimum-wage teenagers, after which engineers lounged in line at the blueprint room. I learned to tell my own engineers, "If you can save $2.00 in direct cost by spending $1.00 in overhead, *spend it!*"

The category of engineering called "industrial engineering" includes a study of accounting by people also educated in engineering. I have found that industrial engineers can often be useful in giving us insight into the economics of the real world. Industrial engineers should be included in our in-house consultants described in Chap. 10.

Auditors

Auditors are accounting police. They search for errors by other accountants in following accounting rules, and they search for fraud by everyone from the company president down. Junior auditors check your time cards for inconsistencies and wish they could check it for truthfulness, but they suffer some restrictions on this side of the iron curtain.

Auditors are used to police the costs of subcontractors who are paid on a cost-plus basis. I hope you will share my relish in a story in which engineers give accountants one in the eye.

Elsewhere I tell of the contract in which my company designed and built portions of a space vehicle simulator for a large aerospace company. I knew that the auditors' practice was to disallow much of the subcontractor's costs, thereby justifying the costs of the auditors themselves to their employer. In negotiating the contract I changed the purchase order boilerplate to allow auditing only of time cards and invoices and to charge time at a fixed price per person-hour. The aerospace company's purchasing department accepted these terms.

The auditors arrived on schedule, with oily smiles, rubbing their hands together, as they asked to see the books and I said no. The conversation was a delight, as their fury rose while I rubbed their greedy noses into the contract. They went home and raised bloody hell, but our work was very good and the contract held and continued. They actually added those time cards and invoices up, down, and sideways and really found about $20 in discrepancies in the $750,000 in total charges. We provided all the office space, courtesy, and coffee they could absorb; no charge.

A public company is required to protect its stockholders from fraud by hiring an independent certified public accountant (C.P.A.; i.e., an accountant licensed by the state) to audit its accounting records ("the books") and its physical inventory and to write an "opinion letter," which is printed in its annual report. Audit by one of the Big Eight C.P.A. firms adds prestige to a company and increases stockholder confidence in the honesty of its managers. C.P.A. companies also provide consulting services and sometimes represent clients in merger and acquisition deals.

Don't Rely on the Big Eight

My robot company had close attention from a partner of our Big Eight firm. He permitted grossly misleading (but "legal") accounting of our R&D expense; helped negotiate a giveaway merger from which he acquired a new client for his firm and a contingent fee (unethical and perhaps illegal), both being clear conflicts of interest; and joined in the coercion which brought me to accept the deal. He later left the C.P.A. firm and got sued by a client.

The chapter on ethics opens with a father-and-son lesson; this one closes with another. "Son, climb this ladder and jump off; I will catch you." Son climbs and jumps; father steps back; son is bruised by the fall. Says father, "Son, this is your main lesson in business. Never trust anyone!"

Part 3

Your Knowledge Base

Technical Knowledge

"The great thing about this information age is that *people* don't have to know *anything*."

Figure 9.1

If your present and future work is to continually refine a small class of existing products without making substantial qualitative improvements, you don't need much new technical knowledge. If your present or hoped-for job is to make substantial qualitative or quantitative improvements in the products with which you deal, the amount of knowledge which might be useful is unlimited. The technical-knowledge base you can use to be successful in inventive or analytical design includes the following.

Your Academic Training

The foundation of your technical-knowledge base is your academic training. Not only did it teach you formal mathematics, science, and some particulars of your specialty, but it should have given you an intuitive insight and understanding of the behavior of those parts of nature with which you deal. Really understanding calculus as a description of nature is a kind of litmus test of whether you actually learned engineering or just learned cookbook procedures for solving standardized problems.

Your academic training may have stopped at the bachelor's-degree level or may have extended to the master's or doctor's level. It may also include any number of single courses given by university extensions, professional societies, and commercial "seminar" companies.

Technicians and engineering

In general, it is a wishful delusion for technicians (including drafters and nonengineer designers) to believe that they can become real engineers by the accumulation of practical experience. The mathematics and science components of formal education are missing from such experience. This is sometimes hard for nonengineers to believe, but it takes a lot of education to know what you don't know.

On the other hand, I have known employed technicians and drafters, some with families, who worked their way through evening college to full academic engineering degrees, and I have much more admiration for them than for those, like myself, who had enough money to attend day college. Let me say further that I can name a number of technicians each of whose value to their company (and as a human being) is greater than that of three engineers I can name.

Continuing Professional Study

After mathematics and science is the art in the field in which you work. (Art is patent language for technology. Using the word "art" saves syllables.) This includes old art, now abandoned, but containing ideas which may not be used in current practice (see Chap. 11, "Libraries and Museums," and Chap. 31, "Minimum Constraint Design"). Knowledge of the current art as practiced in your own organization, in competing organizations, and in research laboratories prevents you from wasting effort in reinventing wheels and gives you a basis for advancing your art.

The principal way you get this knowledge is by study of books and journals in school and for the rest of your working life. Journals include "blue books," which are the scholarly publications of technical societies, and the more easily understood trade journals, many of which are giveaways. I suggest that the most valuable portion of the trade journals is their advertisements. Many of these are very carefully written and illustrated by communication experts because it is of great importance to the advertisers that you understand their products easily and well. The technical merit of the articles often leaves much to be desired, but they are excellent sources of superficial information about fields other than your own. There are scientific and engineering journals of general interest such as *Scientific American, American Scientist,* and some engineering school publications which are rich sources of information and new ideas. There are similar journals for those without scientific or engineering training which are easy to read and introduce new ideas, e.g., *Omni* and *Popular Mechanics.*

Technical society lectures present material before it is published, and you can learn much from conversation with others who attend them. The quality of these lectures varies widely. The lectures are usually printed later on.

New textbooks may be published in your field, some with new information.

Advanced degree?

Should you really study for an advanced degree? Maybe. It costs a great deal of time and a significant sum of money. If you have a family, you must neglect it to an emotionally costly degree. If you have chosen the low-tech branch, the education may be irrelevant to your work and not very helpful to your job progress. A smaller amount of time spent in reading trade journals and catalogs and attending trade shows may pay off much more for you.

On the other hand, if you can afford the time and money as an investment in your future, if you are on or want to be on the high-tech branch, if you really enjoy mathematical engineering, if you want to do advanced research, particularly in a university, or if you want to teach, the advanced degree is essential. (For university work and teaching, "advanced" means Ph.D., the professors' "union card.")

I make no general recommendation because people differ so much. I *do* urge that you spend a lot of time searching your soul and make a careful decision as early in life as you can; do not just drift.

General Knowledge

Knowledge of your own field is not enough. If you would invent in your field, you will find it of enormous value to have knowledge of all other fields and the broadest possible knowledge of nontechnical information and ideas. A classic story in chemistry is of the scientist who conceived the benzene ring. In struggling with his problem he had a fantasy of a snake consisting of a chain of elements; as the snake moved around in his mind, it put its tail in its mouth, and he suddenly realized that this ring structure, the benzene ring, explained the phenomena he was studying, although herpetology was not his business.

In the invention of the mail sorter memory shown in Fig. 2.2 a similar fantasy based on a movie and a popular song was the basis of the code-wheel memory and sensing system used, as described in Chap. 2.

No one knows all the raw materials which now exist, and there is a continual generation of new materials. Many new designs are made possible by the availability of new materials. Therefore it behooves you to watch continually for both old and new materials you might use.

Commercial Components and Vendors

Most product designs include at least some purchased commercial components. It therefore will benefit you to know all about components, not only in your own field but also those developed for other fields, which you might use. The best way I know to keep learning about components and materials is to read trade magazine advertisements and trade magazine new-product releases and to attend trade shows. It is useful to build your own catalog file of components and materials which you think you might use someday. (Most such files are stored by manufacturer's name. I have found that if you store catalogs in library boxes by whatever subject matter headings are most meaningful to you, you will be able to retrieve a box and page through for both ideas and specific products without running back and forth to a manufacturer-indexed file.)

Become acquainted with the manufacturer's directories in your field. They are the best guide to materials, components, and services as well as to products competitive with your own. *Thomas Register* is the most comprehensive I know. I have used it to find sources, ideas, and jobs. *Sweet's* provides combination directories and condensed catalogs in many fields. Other directory companies and magazine publishers print directories which specialize in different fields.

Contract Manufacturers

Sooner or later you will conceive a new product or component which your organization is not qualified to build. There is a large world of specialized contract manufacturers who have equipment or technology which it is not economical for your organization to acquire but who are eager to make unusual components for you. They are represented by sales engineers who are eager to teach you their technologies in the hope that you will eventually buy their services. Free trade literature and sometimes free trial fabrication are offered by them. (See Chap. 10, "Consultants.")

Manufacturing Technologies

Whatever you design must be manufactured if your design is to have value. It therefore behooves you to learn the manufacturing technologies useful for making the products you design or might design. All designers usually have in the back of their minds a technology usable in producing their designs; no one wants to design something that can't be made. If designers are uneducated in manufacturing technology, they will either limit their designs to things they do know how to make or will design products which are needlessly expensive to make. (Design for economical manufacturability is discussed in Chap. 24, "Improving Existing Designs.") Furthermore, knowledge of manufacturing processes will sometimes help you to conceive new components and assemblies to be manufactured. Your knowledge of manufacturing technology should include an understanding of the relationship between quantity and cost for each technique and a knowledge of the many techniques for making models and small preproduction runs. (See Chap. 21, "Quantity Effects on Design.") An excellent source of knowledge of unusual technologies is the catalogs of contract manufacturers. Some catalogs are specialized engineering handbooks.

Trade Shows

Trade shows are rich sources of technical knowledge, particularly of components and materials. You see the real thing, and you can talk to exhibitors who are expert in their products and their application. (More about trade shows appears in Chap. 10, "Consultants.")

10

Consultants

Sources of Consulting Help

There is a vast array of advisers and consultants available to you to help you develop your designs and to help you in business if you become an entrepreneur.

Your friends and acquaintances, including schoolmates you may not have seen for years, form a network of free consultants in an enormous range of subject matter. They will provide you not only with facts and ideas but also with criticisms of what you describe to them; such criticisms can be of great value in keeping you out of trouble if you teach yourself not to be emotionally defensive in the face of criticism.

Within your company, whether large or small, there are people who know things and can do things you cannot do but things which will help you. Your only price is persuading them to interrupt their own work and, if a lot of time is needed, asking your manager to give them some budget. The people themselves will feel some moral pressure to help a fellow employee. Some of these people, such as computer specialists and librarians, are assigned to provide service to others and may be eager to help. In general, they have no conflict of interest with you other than to get more of your budget.

Among the specialized consulting sources of information, guidance, and assistance within your own company may be:

- Advanced mathematical analysts
- People in fields of science and engineering other than your own

- Manufacturing-technology specialists
- Marketers
- Librarians
- Computer specialists
- Commercial artists (for slides, proposal art, etc.)
- Drafters and illustrators

There is an entire industry of paid consultants. These include information experts who can tell you anything that is already known in their field (but will not contribute new ideas) and design consultants who can provide criticism, new design ideas, artistic design ("industrial design"), and mathematical analysis and computer programming beyond your own capability. All the functions listed above are available to small companies from outside consultants. I have benefited from such people repeatedly through my own career, and I work as a consulting inventor and designer at this time.

There is a serious shopping problem in locating a good consultant. A cynical definition of a consultant is "an unemployed engineer." There is an affectation of many solo consultants, including some very good ones, to use the expression "and associates" to seek the prestige of bigness and broad scope. You can get a quick preliminary judgment of a hardware company from published data on credit, age, volume of sales, advertisements, and a catalog examination without subjecting yourself to the pressure of a face-to-face visit from a salesperson, but there is little of such material for consultants.

The ideal consultant for your problem may have started up on his or her own last month after quitting a successful job, may work on a kitchen table, may never have worked as a consultant for anyone else, has no catalog, and probably never heard about Dun & Bradstreet's credit ratings.

On the other hand, there are "universal experts" who will be glad to solve any problem you have since they can always walk away from a failure with your fee in their pockets. They may even believe their own line.

With any consultants you face vendors who want to sell you more and more of their services. If you are lucky, locking them in as permanent assets may be the best thing that can happen to you, but you must be careful not to be sold a lot of time you don't need from people who are better sellers than performers.

If you play it safe and hire a big consulting company, you will deal with a very impressive front representative who will sell you the services of someone else to do the work. You have not yet met the real worker, and his or her work may or may not be equally impressive.

I once made out very well with one of the Big Two by insisting that the impressive front man (who was really very good) do the work himself. He got special permission from his management, charged me a high fee, and proved to be worth many times as much. I also had the reverse experience. For my computer terminal company I went first-class and hired the most prestigious ad agency in town from very impressive front men indeed. I got junk to do the work, raised hell, got a snow job, and fired the bastards.

It is less true in technical consulting than in management consulting, but there are outright crooks and con men out there. I have lost one vice president's job and one entire company to con men. (I have always been able to protect myself personally with good employment contracts, but the waste and destruction were heartbreaking.) Remember that the essence of a con man is that you soon *want* him to hold the family jewels.

How Do You Find a Consultant?

After all these warnings, how do you get a consultant you won't regret?

- Ask people in your purchasing department. They only vaguely understand your technical problem, but they know whom they had hired in the past and left good feelings behind.

- Ask everyone you know in the field who knows a good source. Ask both people inside your company and friends outside. Ask acquaintances at technical society meetings.

- Look up consultant directories in the library.[1,2]

- Telephone, visit, or write to an appropriate technical society office. The society will usually not refer an individual unless the field is very narrow and he or she is unique, but it may give you a list of names of society members in good repute.

- Call a consultant broker. Brokers maintain lists of consultants whom they have screened, at least superficially, and who may have built up good track records with them. The first of those listed will set you up with a phone call first, and you may get your answer over the phone for about $100. I have been on the consultant end of such broker references. The brokers get a reasonable fee for their service. I know two such brokers, and there may be others:

National Consultant Referrals   Teltech Resource Network
187 Calle Magdalena   9855 West 78th Street
Encinitas, Calif. 92024   Eden Prairie, Minn. 55344-3847
Telephone: 619-436-8204   Telephone: 612-829-9000

Some large consulting companies serve as brokers for "staff members," who get paid when they have actual assignments.

- Telephone the head of the appropriate department in any university. Professors work as consultants for conventional consulting fees. If you are young, don't be embarrassed to approach your former lords in this way. If you come with money in your hands, you will be treated with full respect and no condescension. The professors may also refer you on to an expert they know about.

- Use the Yellow Pages in your phone book. It's a blind shotgun approach, but there may be no other. I once found a most valuable hydraulics consultant this way. He was starting a company and supporting it and himself by consulting on the side. I have done exactly the same thing.

To summarize, paid consultants include:

- College professors

- One-person free lances, full-time, part-time, and moonlighters

- Small consulting companies, more or less specialized

- Large consulting companies, usually diversified

I have no reason to believe that any of these categories is better or worse than any other. I have every reason to believe that it is only the individual who does the work who is good or bad for you.

How Do You Qualify a Consultant?

Let us suppose that you have found a consultant who is at least nominally qualified. What do you do to protect yourself from an unfortunate experience?

- Ask for and study a detailed résumé. If it recites degrees and states that the consultant is expert in all aspects of, say, mechanical engineering, get very cautious. My résumé recites every project I ever did and lists all employments and all clients. There are places for a condensed résumé, but a consultant's résumé is his or her sales brochure. It should be complete, and it should not imply experience or capability that the consultant only wishes he or she had. Ethical and responsible consultants will disqualify themselves if they are not skilled in your particular problem subject.

- Ask for and check references. If you can speak to a reference face to face, you will get more frank disclosure than you will on the phone, but I have never heard of anyone who actually does so. The names of individual-person references may not be printed on the résumé to prevent their being bothered too much and becoming hostile.

- Interview. The interview is almost the same as a job interview, but you should also ask for and expect "free samples." It is common for good consultants who know their worth to offer to do the first day's work at no charge to demonstrate their worth. It is also common (but not universal) for the offer to be declined with thanks, but the sincerity of the offer is significant. When I interview, as a consultant, I give the best work I know how. If the interview work is all the client needs, I am the richer for the experience. More often, one assignment leads to another.

What Are the Business Terms?

Prices. As of the fall of 1987, most consulting fees range from $500 to $1000 per day for short periods, plus expenses at actual cost. Juniors who assist a senior consultant cost correspondingly less, and some specialists command much more for short periods of time. For extended engagements, the fees are reduced, since the consultants spend a correspondingly smaller portion of their time doing unpaid marketing. A certain amount of homework to come up to speed on the specifics of the problem, to write a brief report, or just to think is usually thrown in. Serious amounts of homework, whether to study or to produce, are paid for. Fee setting is always to some degree an arm's-length bargaining process.

Location. The consultant works in your office, his or her own office, your customer's office, or anywhere else, depending on the requirements of the assignment and everyone's preference, convenience, and economy.

Terms. Until a satisfactory working relationship has been established, the consultant should be willing to work on a day-by-day basis. "Satisfactory" refers to both technical performance and getting along with the employees and outsiders with whom the consultant must work—the chemistry.

Be sure to specify how frequently the consultant bills you. Consultants with good business experience will be sure that they do not put in more time than you expect them to, but you should share the responsibility of preventing inadvertent overruns. I once had a quarrel with, and lost as a friend, a consultant who soaked in many hours over a long period of time on a project I thought he had stopped work on and then gave me an unexpected bill for the whole thing. (I paid him a compromise sum.)

If the consultant demonstrates that he or she is a long-term asset, you may want to contract for a certain number of hours or days over a certain calendar period (e.g., one day per month for one year) for a designated fee. If the fee is paid on a fixed schedule and the work is on a variable schedule, the fee is a "retainer."

If consultants are on a daily basis, it is unethical for them to leave the project or increase their fees until their work on that project has been completed, but it is permissible for you to terminate their work on whatever basis is specified in the purchase order. I usually suggest that a client issue a purchase order which reads:

> Consulting services, at \$XXX.00 per day plus expenses at actual cost. The work shall be done under the direction of ——— ———. Total charges under this purchase order shall not exceed \$XXXX.00 without further authorization. This order may be canceled at any time without cause.

The "direction of ——— ———" assures that there will be no misunderstanding on the scope or content of the work. The "not exceed" assures the purchasing department that this is no blank check and that this stranger will not submit an ugly overcharge. The "canceled... without cause" assures all concerned that they have not been irreversibly sold a bill of goods. (I myself have had such a cancellation exactly once when a subordinate gave me instructions opposite to his boss's wishes and the overloaded boss blamed me. I have quarreled with clients once or twice, and we have separated without a handshake but with no disagreement about money.)

As a matter of ethics and probably of law (I have never bothered to check), the consultant owes the client both confidentiality and title to any invention made in the field of the assignment (not of the client's total business) during the period of his or her work. Consultants may work for competing companies, but they had better walk a fine line on what they undertake for each and what they disclose to each. So far I have never done so.

Client legal departments and sometimes a macho executive can prevent or destroy a productive consulting relationship with demands for an improper consulting agreement. Usually they ask for assignment of *all* inventions for a prolonged period of time. Usually I just cross out the improper language before signing, and the problem ends. Once, however, I had to fire a client after I had started doing very satisfying work (to both parties) because the president demanded that I sign a document which one of his own directors, a consultant himself, said he would never sign.

Lawyers as Consultants

Lawyers are valuable consultants. Yes, many lawyers really do spend their time suing people, defending criminals, and creating lawsuits, but many others spend their time helping people do constructive work and avoid trouble.

Patent attorneys get you patents, warn off infringers, and try to defend you if you infringe intentionally, inadvertently, or uncertainly (i.e., are you really infringing?). For example, I coined the trade name VECTRON for a robot module producing a vector motion. I used it optimistically without a trademark search. A letter came from an electronics company which used VECTRON as a trademark for certain electronic products and had a clear and early right to the word, telling us to stop. Our respective lawyers worked out an agreement in which we contracted never to use VECTRON for an electronic assembly, the electronics company contracted to let us use it for robot modules, and we both contracted not to go to court about it.

Lawyers are consultants about what laws exist which apply to what you are doing (there are more laws than grasshoppers) and, most important, what these laws mean to you (see Chap. 19). Most laws are written by people who believe that writing clear English is an admission of professional incompetence. Furthermore, the effective meaning of a law is what judges have chosen to make it mean in lawsuits based on the law. Your lawyer will give you an "opinion" about what you may or may not do. The next judge's decision may change everything, so an opinion is the best you can hope for.

If you engage in contract writing for your company, your lawyer will help with the terms. (Your company may have a contract administrator, who in effect is a nondegree lawyer solely in the field of negotiating and performing contracts.) If you start a company, you will live forever with corporation lawyers who are your agents for government relations and contracts of many kinds, who can give you good business advice based on their observation of many other businesses, and who can negotiate some agreements much better than you can. (My present attorney has done wonders for me.) However, watch what is going on; you know more details of your situation than your lawyer does.

I have spent a lot of money on legal fees, and I feel that I have gotten excellent value (except for one lawyer who, with the aid of an ignorant judge, a better opposing lawyer, and a technically competent consultant who was a perjurer for hire on the other side, lost me a $40,000 lawsuit).

In your personal life you will encounter legal problems, some of which may be constructive and happy, like buying a house. I advise you to have a lawyer and consult him or her early and often. It's well worth the money.

Vendors as Free Consultants

Vendor sales engineers and even salespeople are a most valuable class of truly free consultant. They are not charity. They are sent to you by a vendor in the hope that the assistance they provide will encourage you to buy from that vendor. They teach you knowledge of the vendor's products, services, and organization, and they provide application guidance for using the vendor's products to solve your problems. The better ones will help you adapt your design so that their products will help you solve its problems and will even help with engineering problems unrelated to their products. (This is called "consultative selling" and can bring prestige to the sales engineers' company.) You are expected to make allowance for their competitive bias, and you are free, legally and ethically, to get such service from competing vendors. Vendors may give you detailed written proposals, including special designs, for using their products to solve your problems as well, of course, as price quotations. (I consider it unethical, as a buyer, to disclose the contents of one vendor's proposal to another vendor in the hope of merging the best of both proposals into a single proposal. After an award has been made, however, unless there has been a confidential-disclosure agreement, there is no such inhibition.)

Vendor consulting extends to seminars, courses, and books in the vendor's technology. All these are marketing investments by the vendors, so you may be sure that they try hard, in their own interest, to be sure that you are well taught. Some courses and books are charged for, largely as a mechanism to limit attendance and acquisition to serious prospective customers. Some catalogs and handbooks are written as textbooks as well as as a display of wares, and their language is usually edited by serious advertising writers to make it clear and unambiguous. (I am not referring to fliers showing a pretty girl sitting on the product to suggest that she comes with it.)

Some vendors will provide free samples (resistors, yes; 100-hp motors, no), and others will provide loaners for test (perhaps a 100-hp motor). A vendor introducing a new machine tool or other expensive product may place one free in the plant of a prestigious customer, asking only that it be given a serious workout in actual service and that prospective customers (including the customer's competitors) be permitted to visit the installation and ask about its performance.

I should like to suggest, as a matter of ethics, that you not bleed vendors of such assistance unless they have a fair chance of making a sale. The closer the relationship comes to serious business, the more justified you are in asking vendors to invest serious cost in proposals, presentations, demonstrations, and the like.

The better your track record with vendors, the more they will invest in helping you. My biggest and best customer in the robot business was IBM. Its people really got ethics lessons and really obeyed them. They would not let us spend any effort on a sale until they had an authorized budget. Guess how we responded when we did get the opportunity.

Many companies provide videotapes as free loaners to supplement their catalogs. Such tapes can be most informative as well as very easy to understand. I provided them, in all formats, for my robots, and they were asked for all over the United States. Some libraries also lend videotapes on a great range of subjects more or less devoid of commercial sales pitches.

It is common for vendors to provide seminars for engineers, either within large companies or in hotel space. These seminars are sales pitches, but as with all good advertising they sell by educating and are well worth attending.

Trade Show Consulting

A most valuable form of free vendor consulting is found at trade shows. Not only can you examine actual hardware products and see them demonstrated, but you can consult on the spot with sales engineers, sometimes including the vendor's vice president for engineering. Immense amounts of application engineering are carried out on the display floor, in hospitality suites, and at lunch and dinner at trade shows. If you are lucky, you may learn valuable information from the technical papers usually presented as a parallel activity to the product exhibits.

You can also collect and order pounds of catalogs, sightsee, and lounge in the hotel bars and hospitality suites if you feel that your career success is best served by doing so.

Government Technology Transfer

Government agencies have technology transfer departments which are directed by Congress to provide the results of unclassified research to industry. They are free consultants. Agency personnel are motivated to do so in order to help justify their own jobs. Such agencies include the NBS, various NASA headquarters and laboratories, and specialized laboratories.

In this discussion I have spoken of technical consultants. There is a large class of management consultants that I will not discuss in this book.

REFERENCES

1. *ACEC Directory,* American Consulting Engineering Council, 1015 Fifteenth Street, N.W., Suite 802, Washington, D.C. 20005. Telephone: 202-347-7474.
2. *Consultants and Consulting Organizations Directory,* 8th ed., Gale Research Company, Book Tower, Detroit, Mich. 48226, 1988.

Libraries and Museums

Libraries

When you have a specific assignment, you need to do focused research on materials, components, and the state of the art as evolved by others. Most of this you do in libraries.

Locations

There is a enormous range of libraries accessible to you if you will use them. These include:

- Your own collection
- Your colleagues' collections
- Your company's library
- University and technical school libraries
- The technical room of your public library
- Specialized libraries in technical societies and other institutions
- Government agency and laboratory libraries

There may be surprisingly rich libraries in your own organization. In many organizations there are specialized groups such as material specialists, mathematicians, and so on, each with a personal library. They are usually glad to help out a member of a different group because doing so makes them feel good and important and because it may transfer some money to their own accounts. Remember that your marketing department has a library on your competitors and on other useful material which may or may not be "engineering." In my aerospace company there was even a group which collected appropriate Russian publications.

Your marketers and salespeople are the G2 (intelligence, in the military sense) of your company and are probably eager to help you. Furthermore, you can make some success points with them if you tell them what you have learned about a competitor in a trade show or technical meeting and give them some catalogs you have picked up for their library.

Contents

Libraries contain the following kinds of material:

- Reference books.
- Purchasing directories (I have started many projects with *Thomas Register*).
- Textbooks.
- Encyclopedias.
- Periodicals.
- Videotapes.
- A wide variety of materials such as photograph collections, theses, monographs, etc.
- Computer terminals with access to computer databases all over the United States. (See Ref. 3.)
- Collections of abstracts of technical papers.
- Indexes referring you to other publications such as *Readers' Guide to Periodical Literature* and *Engineering Index*.
- References to foreign publications and translation services. The development of foreign countries in the past two decades makes this an increasingly important class of sources.

References 3 and 5 are surveys of such information sources.

Each library has its own storage, filing, indexing, and cross-reference system. Some of these are on computers, and you may be permitted to use the computers yourself.

Patent Office search room

There is a unique library in the United States which is sadly neglected by design engineers. It is the U.S. Patent Office search room in

Washington. In it are copies of all United States patents, filed and cross-filed in accordance with an elaborate but easily understood manual of classification. The stacks are open to the public, and there are even special fixtures on the reading-room tables to help you manipulate the patent copies. There is a staff of highly trained and cooperative librarians whose sole function is to help you find the patents you are seeking.

Why is this library useful to design engineers as well as to the patent searchers who are its principal users? The theory of the United States patent system is that the government exchanges a limited monopoly to inventors in exchange for their teaching the public the art which they have invented. This teaching takes place in the specifications of their patents, and if they do not make a thorough disclosure of their new art, their patents are invalid. These patents are a mine of ideas, many never exploited in the marketplace for reasons unrelated to technical merit. They are there for you to learn from. (Chapter 2, "Inventing," has more on the patent system.)

A trip to the Patent Office requires a cash expenditure by your company over and above your own salary. It must be sold to your management. Put this on the list of applications for your Golden Art of persuasion, as discussed in Chap. 4.

Librarians

A major asset of a library is its librarians. It is *not* the primary function of a librarian to look stern and say "Shhh." Librarians get their job satisfaction from helping you find material, and they get the same feeling of pleasure in making a search hit as you do in coming up with a good idea. They are professionally trained, have college degrees in library science, and have incredible amounts of knowledge of what material there is, where it is, and how to get it. They can arrange for you to visit other libraries and to borrow material from them. Some librarians will help you build a bibliography in the subject of interest to you. Librarians are eager to help and enormously helpful; use them!

An online search is the querying of a database by a computer and the printing out of those items in the database which are called for by the querying computer. Such database searches are provided by some libraries and may be available through your own company computers. For example, among the data bases which can be accessed is *Engineering Index.*

The databases in and through your own personal computer are a part of your library. The use of such computer databases is so extensively covered elsewhere that I only mention it here to be complete but will not discuss it further.

Your personal library

You will build your personal library throughout your career. If you do a particularly good job, it will enhance your reputation; others will consult you and thus contribute to your success in a second way. I'd like to make a few general suggestions about your library.

Many commercial catalogs are sources of technical information as well as the vendors' product data. They are usefully filed among textbooks and copies of technical papers. For this reason they are most usefully filed by subject rather than by manufacturer's name. It may be desirable to secure two or more copies of a catalog or to cut up a catalog in order to file each portion of the material in the appropriate subject location.

A most useful piece of library hardware is the document box. It is a cardboard box about 9 by 12 by 3 in into which you place booklets, clippings, etc., which are not self-supporting like a book. Adhesive labels identify the contents. I have built large catalog files with such boxes.

If you are a disciplined and organized labeler, the standard Pendaflex file permits very fine subdivision with easy access of your reference papers and record papers.

Some people build libraries of clippings and materials they might use someday. I know a very fine medical researcher who saves cartoons and who always has apt cartoons to illustrate the many papers he delivers with technical slides. I wish I had done this so I could use the material to enliven this book. (See Fig. 9.1.)

Some people save technical journals. I am skeptical of the volume-to-utility ratio, particularly since journals are easily accessed in libraries.

Technology Museums

There are technology museums in the United States and Europe which provide a fascinating presentation of the history of technology and great exercise for the visitors' imagination. You can spend useful hours in front of their exhibits imagining the mental processes of the designers, limited by the knowledge and facilities they then had, to understand why they made the designs they did. I have done so, and I believe that this exercise not only was of tremendous interest but was useful in my own training. You get an insight into the nature of devices you have always taken for granted by seeing how they evolved from primitive versions.

I speak of true technology museums. These should not be confused with the science museums which attempt to teach physics to children and succeed only in providing toys.

Here is a list of technology museums I recommend:

Smithsonian Institution Washington, D.C.	Transportation Museum Lucerne, Switzerland
Ford Museum Dearborn (Detroit), Michigan	Museum of Technology Vienna, Austria
Science Museum London, England	Deutsches Museum Munich, Germany
International Museum of Horology La Chaux-de-Fonds, Switzerland	Széchenyi Manor Museum Sopron, Hungary

A complete list can be found in Refs. 1, 2, and 4. The European museums make wonderful stops on a European vacation. However, the exhibit labels are in the local language and may be frustrating.

I should warn you that these places are addictive. It is much easier to go in than to get out.

REFERENCES

1. Hudson, Kenneth, and Ann Nichols (eds.): *The Directory of World Museums*, 2d ed., Facts on File, New York, 1981.
2. *Museums of the World*, 3d ed., K. G. Saur, New York, 1981.
3. *The North American Online Directory, 1987*, R. R. Bowker, Box 766, New York, N.Y. 10011. Telephone: 212-337-6934 or 800-521-8100.
4. *The Official Museum Directory*, American Association of Museums, Washington, National Register Pub. Co., Wilmette, Ill. Revised annually.
5. Schenk, Margaret T., and James K. Webster: *What Every Engineer Should Know about Engineering Information*, 1st ed., Marcel Dekker, New York, 1984.
6. *Thomas Register of American Manufacturers*, Thomas Publishing Company, New York. Revised annually.

Chapter 12

Your Company

In order to be as successful as you can be in your company (that is, to produce designs which are appropriate, which are accepted, and which prove useful to your company) you should understand your company itself. What should you know, and where can you find out?

Engineering Department

Your engineering department has written and unwritten policies, ground rules, capabilities, limitations, and controlling personalities. It has offices, laboratories, fabricating shops, computer systems, and staffs of both hardware- and paper-making technicians. There are standard drafting conventions, paper sizes, and dimensional units. The United States is in the process of partial metrication, but the degree varies from company to company and within companies. You will save yourself a lot of trouble by conforming to the practices of your company unless you care to generate feuds and crusades. Save your energy for those crusades that you carefully select to be worth waging.

I once did a consulting job in a European country. In order to be a good fellow I tried expressing myself in metric units and estimated that something "was about 10 cm long." The chief engineer turned on me in a rage and shouted, "Centimeters are for carpenters; engineers use millimeters!"

Manufacturing Department

Even more varied than the policies, capabilities, and personalities of your engineering department are those of your manufacturing department. Life will be easier for you and more of your work will be accepted if you design products which can be made in your own factory by techniques it already practices.

It is a matter of judgment how far you compromise what you feel is an optimum design in order to conform with your own plant's capabilities and preferences. (Remember that you have judgments and other people have prejudices.) You should properly pioneer new manufacturing technology and the use of vendors with specialized technology that your company does not have, but you should do so only when you cannot adapt your designs to your factory without serious loss of design value.

This is more important in some plants than in others. Some companies make very complex products with almost no capital equipment by buying all fabrication from outside vendors and doing only assembly in house. Others feel strongly that they want to produce their own products in house, either for the feeling of control or for other reasons or feelings. You will be a more effective and productive designer if you understand the manufacturing policies of your company and work to accommodate them.

Here again you are always free to start a crusade, but the energy in waging crusades is energy taken from your design career. When I was younger and more foolish, I banged my head against a lot of stone walls. I would be better off today if I had not.

Marketing Department

Your organization sells something to get the money which pays you. It may sell grants, research projects and reports, computer software, small quantities of specially designed hardware, standard capital equipment, or mass-produced consumer products. The reasons for its choice of marketing arena in which to sell may be historical, rational or irrational, or the personal taste of managers whether with good reasons or with bad. Sometimes an organization, such as a military products company deciding that it would like to enter a commercial market, is in transition.

There is a very high probability that different people in the organization have different opinions about the market it should be in. Conflicts may arise of which you should try to be aware and in which you may participate, voluntarily or not.

Your product proposals and designs should be appropriate for the market to which they will be addressed. A coffee maker for a long-range bombing plane should be different from a coffee maker sold in a low-price department store, and both should be different from a coffee maker sold in a high-price gourmet specialty shop.

Note that the marketing *policies* of your company are based at least in part on the marketing *capabilities* of your company. (The relationship is chicken-and-egg.) If your company now sells research contracts to the government and you have a great idea for a mass-distribution product for consumers, be aware that setting up the salespeople, representatives, supervisors, warehousing, advertising, credit system, and so on for your proposed product is not a trivial undertaking. Your marketing department also is made up of personalities with their own motives and prejudices, which they may not clearly announce for all to understand even if they really understand them themselves. (You and I know that only engineers make rational judgments undistorted by emotion.)

Unwritten Ground Rules

There are other aspects of your company which you should try to find out. There may be unwritten and sometimes unspoken ground rules:

- "We like to make our products of metal and not of plastic."
- "We like to make our products of plastic and not of metal."
- "We like easily maintained products."
- "We like disposable products."
- "We like products which can be maintained only by our own service organization so that we can sell maintenance contracts."
- "We like products which require a certain manufacturing process which is available only from the president's brother-in-law."
- "We have a department in the factory which must be supported because we want to maintain jobs for certain people."
- "We have a troublesome department in the factory which we would like to close down."

The number of such unwritten ground rules can be legion, and the better you know them the more successful your designs will be.

Company Culture

The phrase "company culture" refers to the complex of unwritten ground rules within which the employees are accustomed and expected to operate. It may vary from complete indifference to anything except doing assigned work to a complex set of unwritten rules which may include dress codes, manners, who associates with whom, and the like. The military is a limiting case in the spectrum of company cultures.

You should consciously learn what your company culture is, decide whether you are willing to live with it, adapt your own practices for success in that culture, or else start looking around.

Attitudes toward innovation

A most important aspect of your company culture is the attitude toward innovation of the different managers. (Chapter 2 describes the innovation index. This is the receptivity of a person to new ideas not originated by himself or herself.) If you are working for people with low II rankings, don't frustrate yourself and get a bad name by bothering them with a lot of "crazy new ideas." If you are working for people with high II ratings and if you enjoy and are good at invention, you will have a wonderful time and a warm reception by showing your ideas to your managers.

Your company or your immediate part of it may have any of several attitudes toward the strictness with which your time is spent as assigned. There are organizations with the "anyhow theory of economics," which says that the engineers get paid anyhow, so it doesn't very much matter what they do but that discretionary expenditures of real cash money must be kept very low. I have seen engineers carrying drawings back and forth to the blueprint room because some clever efficiency expert had devised a cost reduction by eliminating errand runners. On the other hand, in such an organization it may be easy to bootleg time to develop a new proposal to your management.

Sources of Information

A source of information about your company is the annual report it sends to the stockholders. In it your company president and board of directors explain to their bosses what they think the company is all about, how well it is doing, and what they hope to do. (Remember that they are defending their jobs and also selling stock, so use a grain of salt in your reading.) You will probably be more successful if your designs and proposals match the thinking expressed in the report. (Don't be embarrassed to ask for a copy; the officers are usually proud of it and want it spread around. If you are anxious anyway, ask stockbrokers to get it for you. They will be glad to do so because they might get to sell you some stock and they have no need to mention your name when they ask for the report.)

If you work for a public company (one whose stock is held by the public and can be freely traded), there is a great deal of information about the company, its officers, and its industry in reference books called *Moody's* and *Standard & Poor's*. You will find these in brokers' offices (just walk in and ask) and in your public library. These books are written primarily for investors but are just as useful for you.

If you buy a few shares of your company's stock for a few hundred dollars (any stockbroker will serve you), the company will send you a copy of the annual report every year without your asking, and furthermore you will be invited to its annual stockholders' meeting, where you can see the top brass report and defend themselves to people they can't discipline or fire. If you have the guts, you can address them directly yourself.

Many brokerage companies have analysts who study companies and write reports which are used by the brokers to encourage their clients to buy or sell the corresponding stocks. These reports are usually open to the public in the brokers' offices. I make no general comment on the accuracy of these reports, but they may provide information and ideas not obtainable elsewhere.

When your company first went public and whenever it goes back to the public to sell more stock or bonds, it must file a very detailed report (typically form S1) with the SEC. The officers and auditors are subject to severe penalties if the report is incomplete or misleading. (The annual report must also be filed, under similar rules.) These reports are bulky and not easily found except in an SEC office. However, a condensed version of the report is the "prospectus," which a broker must give to each prospective first buyer of the new stock or bond issue. Ask any broker for the last prospectus.

Other sources of information about your company are its organization charts (some of which may not be easy to get), its commercial advertising in both trade and financial journals (the two kinds of ads are different), and the sales brochures and catalogs it sends to its customers.

Every company management sends messages to its employees through:

- House organs (company newspapers)
- Meetings and talks by managers
- Printed announcements
- Company socials (picnics, Christmas parties, and the like) in which managers mix and talk with others than the adjacent names on the organization chart
- Unofficial social life (dinner parties, etc.) in which employees associate

The more you know about your company, the more successful you will be operating within it.

Your Market

If you are to be successful, you will design to suit your market. What are the attributes of this market?

Culture of Your Market

In addition to the expressed requirements of written specifications, standards, and codes (Chap. 19), there are the unwritten requirements of the culture, conventions, and fashions of the market to which you are selling. This is true not only in the field of consumer products but also in the fields of technical and capital goods. You must also know the individual people and groups to which you offer your ideas and tune the character of those ideas to suit the taste of the customer if your ideas are to be accepted.

Receptivity to innovation (innovation index) varies from market to market. The electronics industry, particularly the computer industry, has always been eager for new and better products. The capital equipment market in the metalworking industry and, in particular, in the machine tool industry is the reverse. We all have our own opinions of the automobile industry. The military and other government markets vary from one extreme to the other, so you must know the particular part of the government to which you are selling and, if possible, the individual decision makers who affect you.

The women's shoe industry is the worst I know. See the story of the Department of Commerce meeting to "save the shoe industry" (Chap. 2).

When I was a proposal writer in the space business, we regularly received specifications from NASA which called for the use of "standard, off-the-shelf, space hardware" as components of the satellites which we proposed. This was in 1960, when there were only a few satellites in space! On the other hand, there are portions of the military which are aggressively eager to consider serious innovation.

For another example, it took over 20 years to persuade the machine tool industry that numerical control was a good thing and to replace most tracer controls with numerical controls. Today, of course, no machine tool manufacturer would even consider a tracer control regardless of the advantage its use might offer in a particular case.

An example of fashion in industry is the use of digital displays instead of moving-pointer (analog) displays in automobiles, appliances, instruments, and machines. A friend of mine is a senior manager in a very high-tech company. He told me that he was looking for an automobile with only digital displays on its dashboard. Since he was a leading advocate of computerization in his engineering department, I asked him if he urged the use of graphic displays in his computers. He enthusiastically said, "Yes." I asked, "Aren't graphic displays analog displays converted from digital data for easier understanding by humans?" He got angry and walked away.

Types of toys come into fashion and go out of fashion as rapidly as fashions in clothing. The same is true of fashions in military hardware, airport configurations, and other products in which a change in fashion has little relation to developments in technology.

Classification of Markets

One can classify markets in many ways, but I think the following will be useful in your own thinking as a design engineer.

In-house

There may be an in-house market in your own company for manufacturing machinery, R&D instruments, and brick-and-mortar projects. Often your in-house market will tolerate aesthetic defects such as toolmarks, weld burns, scratched paint, and other visible residues of debugging changes. On the other hand, it may require conformance with in-house standard maintenance parts, tools, and practices.

Competitive bid

There is a market ruled by competitive bid to detailed specification. Most government purchases and many industrial purchases are made on this basis. Here the challenge to you is to conform to the specification with an absolute minimum of cost required by your design. It is sometimes permissible to make unsolicited proposals to customers for

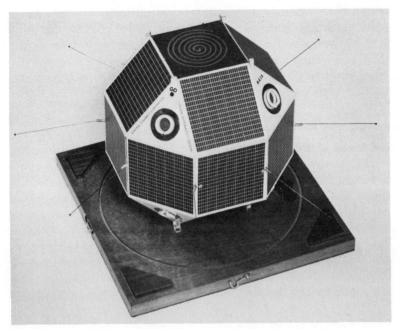

Figure 13.1 Proposal model of geodetic satellite. (*Courtesy of General Dynamics Corporation.*)

changes in a specification which will give them features they did not think they could get or afford, remove features which are unnecessarily expensive at the benefit of reduced price, and incidentally reduce your own competitive price in doing so. Ground rules may or may not permit such unsolicited responses. Sometimes one or more vendors help to prepare the specification with proposals and meetings. For major military programs several competing vendors may be awarded contracts to develop competing proposals. (See Chap. 19, "Specifications, Codes, Standards, Contracts, Laws, and the Law.")

There is a similar market for competitive proposals to a broad specification. The expression "request for proposal (RFP)" is used. Here you are faced with a competitive opportunity for a combination of best design and lowest cost. You must exercise judgment about the taste of your customers as they weigh the value of features against cost. If you have innovative ideas to propose, you should judge their innovation index (receptivity to innovation; see Chap. 2).

(The word "judgment" appears repeatedly in this book. Judgment is making decisions on values which cannot be computed. Your personal success will depend to a large degree on how good your judgment is and becomes.)

Such a response may be good engineering yet fail to win. I once responded to a NASA request for a geodetic-satellite quotation with an unsolicited variation which I thought was better. NASA not only agreed that it was a better design but canceled its request for quotation (RFQ) and issued a new RFQ incorporating my drawings. A different company won the award on price. (Figure 13.1 shows the model we submitted with our proposal.)

Original equipment manufacturer

The next class of market is the original equipment manufacturer (OEM). This is the market for components which are sold to other manufacturers for incorporation into their products. Raw materials, transistors, motors, and hydraulic cylinders are examples of OEM products.

(I know of no distinction between a "component" and a "system" other than that every company buys components and sells systems. To a resistor manufacturer a resistor is a complex system with components of wire, insulators, paint, etc. To an admiral, an aircraft carrier is merely a component of a fleet.)

Capital equipment

Capital equipment includes manufacturing machinery, trucks, earthmoving machines, buildings, laboratory instruments, and any other product which is not consumed in a short time or is not resold to customers.

Consumables

Consumables are products which get used up in a short time by consumers, commercial companies, or manufacturing companies. A photocopy machine is capital equipment; the blank paper it uses is a consumable.

Consumer products

The next class of market is consumer products, which include durable products such as can openers, stereo equipment, automobiles, house furnishings, etc., and consumables such as food.

Classification by user competence

Another way to classify markets is by the education and skill of the user in the product's technology. Educated-user products include electronic instruments and other technical devices which will be used by trained personnel. Uneducated-user products include most consumer products. An educated customer like the Army may buy for uneducated users like private soldiers. The difference affects permissible complexity of operating controls and displays, skill required to use the product, simplicity of language and amount of detail in the instructions, and maintenance practices and procedures chosen by the designer (e.g., fixable versus disposable).

The human environment for different products has extreme variations. In your product's operating environment will your product be used by truck drivers, housewives, or skilled technicians? In your product's maintenance environment will the people be trained? Untrained? Equipped with an inventory of spare parts? Equipped with good maintenance tools? Disciplined and goodwilled or undisciplined and malicious?

Your product may be subject to maintenance only by trained maintenance people employed by your own organization or by distributors for your organization. In this case you can assume that the maintainer is unlikely to do damage, understands the machine, is equipped with special tools, and is capable of rather educated diagnostic thinking. Next there is the class of professional maintainers such as garage mechanics who are more or less trained and more or less well equipped. Third, there is maintenance by the user who may be highly qualified, as in the case of electronic technicians in research laboratories, or may be highly unqualified, as in the case of users who include the uneducated and unskilled.

Related to the maintenance problem is the problem of the installation and maintenance instruction manual. The manual must be written to suit the class of customer. You may have the services of a technical writer to help prepare such a manual, but you must tell the writer what to put into it.

As the designer you have the choice of designing for special maintenance tools, which makes your design life easier, or working very hard to make maintenance easy with standard tools. In some modern electronically controlled products it is the practice to design in very complex diagnostic devices and circuits which display to the maintenance person the nature of the fault and the action required. It is sometimes very desirable to put conspicuous displays of need for supplies, such as paper in a duplicating machine, so that the user is not tempted to fix a nonexistent fault and thereby create a fault. The medical community uses the word "iatrogenic" to describe disorders produced by doctors attempting to cure other disorders. (That's a dandy word to show off with.)

The establishment and availability of hot-line operating and maintenance system service permits a degree of complexity in operation and maintenance which might not otherwise be permissible.

I once lost a robot sale because, the day before, a mechanic had smashed a die by slamming it into position with an iron bar instead of carefully levering it into position. The customer had no faith in the ability of his people to care for our machine. I have been warned that my machines should be sabotage-resistant in certain plants. Products exposed to the public, who neither are employed by the product's owner nor own the product themselves, are subject to vandalism. Consider the design of the public telephone, which is remarkably vandalism-resistant. I once met a very capable engineer employed by one of the Reno gambling casinos who was full of stories of slot machines smashed by drunks and losers. (See Chap. 22, "Reliability and Maintenance.")

Military versus civilian

Military and other government customers differ in many ways from industrial and commercial customers. Military specifications (MIL specs) cover the product, its components, shipping, general performance such a radio-frequency interference, and your own inspection practices. There are usually requirements, absent from consumer products, such as shock and vibration resistance, fungus resistance, and many others. (See Chap. 19, "Specifications, Codes, Standards, Contracts, Laws, and the Law.") Many companies choose to make only military products, many choose to make no military products, and some undertake the administrative complexities of making both.

Foreign versus domestic

Foreign markets differ from domestic markets in a number of ways. The applicable codes and specifications are different. Local practice probably uses metric dimensions, so that fasteners and other replacement components which must be obtained locally and maintenance tools must be metric. Instruction and maintenance manuals must be in the language of the customers. Labels must use language-free standard symbols. Packaging must be suitable for overseas shipment and must meet the requirements of the commerce and customs government authorities in both shipping and receiving countries. (See Chap. 32, "Design for Packaging and Shipping.")

There are other ways to look at your market. The physical environment for your product may be anything from friendly to hostile. An electronic instrument for a research laboratory does not face operating conditions of shock, vibration, humidity, fungus, rain, heat and cold, and rough handling, but a radio set for a company of marines does; a machine in a food plant must withstand severe washing and must not leak lubricants, a medical instrument must withstand sterilization and strange chemicals like blood, and an oceanographic instrument must withstand enormous pressure and the corrosiveness of seawater.

Market size

The size of your market affects the nature of your design primarily by dictating the amount of nonrecurring cost which may be invested in order to minimize recurring or unit cost. There are many differences between the design of an automobile sold by the millions per year and the design of a large newspaper printing press sold once or twice per year.

My final suggestion on the subject of designing for your market is that you discuss the product in detail with your marketing and sales departments before you get beyond the initial design. Then it will be most acceptable to them and to the market as they understand it, and you won't have to do expensive and exasperating redesign when they get to criticize it.

Your Competition

Competition in the World

Competition is a part of our lives. In school we competed for both academic and athletic standing. We play games like tennis or chess in which we compete for the sake of competition. We compete for dates and mates.

In some cultures competition is bad manners. In ours competitiveness is a major virtue but is supposed to be practiced with good manners. On the playing field this is called sportsmanship, is supposed to be practiced by gentlemen and ladies, and sometimes is. In industry sportsmanship reduces to rudimentary courtesy. In criticism between architects and between academics it sometimes reduces to savagery. Espionage, planting disinformation, and dirty tricks are not unknown. (See Chap. 5, "Your Ethics.")

Your Competitors

As a design engineer you compete with other design engineers both inside your company and outside it. If your company sells a product, then you, as part of your company, are competing with other vendors attempting to sell their product in place of yours.

You may compete with an in-house department of one of your customers. One of my clients makes active dampers for automobiles. Once they made a presentation to major auto manufacturer H but were told that the customer had an in-house group working on a similar product and that the group's design was favored because it was in house. My client pulled a working model from his briefcase and presented it. The model put him ahead of the customer's in-house group, and he won a contract as a result. His competitive victory was based in part on the technical merits of his design, but he won largely because his extra energy and diligence had moved his program farther and faster than his competitor's program. We all compete with the calendar and the clock.

Within your own organization you compete with others for budgets (including raises), for facilities, for helpers, for rank and position, for desirable assignments, and for the selection of competing ideas.

If you work in a government military laboratory, your competitors are called Russians.

If you work on almost anything, some of your competitors are called Japanese.

If you work in academic research, your competitors are other researchers racing you to publication, recognition of priority, prestige, promotions, political position, and grants.

If you work in a large corporation, there is competition for budget and assignments between the R&D departments of the divisions and the corporate central R&D department.

If you work in a corporation, you are faced with the competition of contract R&D companies and universities that propose to your management that they can do better work for less money than you do.

If you work in almost any kind of organization, you help it compete with other organizations for sales and contracts.

What Wins?

In football the components of winning are touchdowns and field goals. What wins for the design engineer?

The first winner is design merit, whose elements are discussed throughout this book.

The second winner is speed—getting there first.

The third winner is persuasion—making your customers believe that buying from you is best for them. Persuading them that your design has the most merit is part, but only part, of that persuasion. I keep referring you to Chap. 4, "Persuasion: The Golden Art."

The fourth winner is prediction, one of the skills I encouraged you to develop earlier in this book. What are the competitors thinking? What will they do? Which way will they turn? How do their minds work? If you know a lot about them and their circumstances, you have a good chance to create a design strategy which will beat them. That's how generals win battles. (See Chap. 18, "Prediction as a Design Process.")

Topics in
Design Engineering

Simplicity

Everyone is in favor of simplicity in design just as everyone is in favor of reliability, cleanliness, patriotism, and motherhood. Like beauty, people can't explain it, but they know it when they see it.

Thirteen Ways to Keep It Simple

1. Use as few modes of power as the requirements permit. You can operate different parts of the same machine with ac electricity and dc electricity, both at various voltages, pneumatics of different pressures, hydraulics of different pressures and fluids, internal combustion engines, and steam. Each may be ideal for its particular load, but would the overall machine and its customer be better off if everything operated from dc electricity at a single voltage?

2. Use as few modes of control as the requirements permit. You could combine digital electronics, analog electronics, analog pneumatics, relays and semiconductors, pneumatically and hydraulically operated valves, and mechanical linkages, but a single microprocessor with output amplifiers feeding electric motors and solenoids and fed by transducers with electrical signals is hard to beat these days.

3. Use as few kinds of materials and components as the requirements permit.

4. On control panels, group displays with corresponding controls. Separate groups of controls. Array controls to model their effects; e.g., put the up button over the down button. Follow conventions. Read Chap. 29, "The Human Interface."

5. Form the habit of critically examining machines, buildings, or-

ganizations, furniture, and everything you see for simplicity, aesthetics, economy, and functionality.

6. Combine two or more parts into a single part. Now the going gets tougher because, depending on how good you are, this rule can result in either simplicity or complexity.

7. In electrical circuits, particularly, struggle very hard to reduce the parts count.

8. In electrical systems use a minimum variety of conductors, cables, switches, and other components. In pneumatic and hydraulic systems use a minimum variety of tubes, valves, cylinders, and other components.

9. Use a minimum variety of fasteners and other hardware. Standardize on one type of screwhead. Design for a minimum variety of maintenance tools.

10. Provide easy access to components for trouble diagnosis and replacement both in manufacturing and in field maintenance. If you stuff a bunch of electrical parts and wires into a sharp-edged, deep, inaccessible hole, your customers will curse you forever.

11. Bells and whistles: are they all worth putting in? Should some of them be customer options?

12. Form the habit of critically examining machines, buildings, organizations, furniture, and everything you see for simplicity, aesthetics, economy, and functionality. (I know that I repeat myself on this point, but I think the repetition is useful.)

13. Think. Think very hard about how you can make your design simpler. If you really want to and if you really try, you will make it simpler.

If I may be permitted another bit of pontification, in science, art, speech, writing, organization, and engineering, complexity is easy; simplicity is the result of great effort.

Simplicity and Advanced Technology

Simplicity is *not* the elimination of advanced technology. Of course, advanced-technology products may be more complicated than elementary-technology products, but they may provide more function also.

This is a good place to introduce the word "sophisticated." *The Random House Dictionary* includes two concepts, among others, in its definitions: "complex or intricate" and "worldly-wise."* A simple design is worldly-wise; a complicated design is complex or intricate.

The word "sophisticated" has added some new meanings in recent

---

*By permission from *The Random House Dictionary of the English Language, Second Edition, Unabridged,* copyright 1987, by Random House, Inc.

years. An advanced-technology product is said to be sophisticated: e.g., "a sophisticated radar." A person with advanced education, training, or experience is said to be sophisticated: e.g., "a sophisticated user." In the law of finance, a "sophisticated investor" is experienced and understands about investments and their risks.

More on the *Challenger* Fiasco

One of the worst designs I have ever seen, both from its complexity (lack of simplicity) and for its failure to perform, is the solid-rocket field joint and seal of the *Challenger*.

Figure 15.1 is taken from the *Report to the President by the Presidential Commission on the Space Shuttle Challenger Accident* (the Rogers report). The mechanical joint is an annular tongue and clevis 12 ft in diameter with 180 clevis pins retained by an external band! The seal is two O rings in series, each oriented so that its groove-depth and mating-face tolerances are tolerances on a 12-ft diameter. Since the walls are only ½ in thick, the cylinders spring out of round and a hydraulic jack is provided for field assembly to squeeze them close

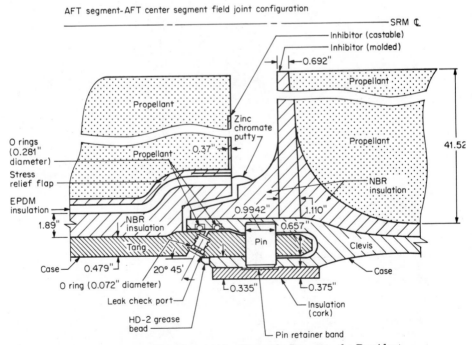

Figure 15.1 *Challenger* joint-and-seal assembly. (*From the* Report to the President by the United States Presidential Commission on the Space Shuttle Challenger Accident, *Government Printing Office.*)

enough to round so that they will go together. A leak check port permits injection of compressed air between the O rings to test for leakage. This air moves the inner O ring to the wrong end of its groove so that it must move back when working pressure is applied (which should do little harm), but the air which does leak past the inner O ring tunnels through the heat-insulating putty and establishes a path for the hot rocket gas to reach the O rings. The designers' ignorance of how an O ring works was described in Chap. 1. This design has now been improved: it will now have three O rings in series!

After I finished denouncing this design, I asked myself whether I could do any better. Figure 15.2 shows the results of about an hour's work on this exercise. There is a single O ring in a groove of standard handbook dimensions. The groove and mating surface are on the faces of the cylinders rather than on the diameter, so machining tolerances affecting squeeze are on a ¼-in dimension rather than on a 12-ft dimension. Mechanical coupling is done with a conventional coupling

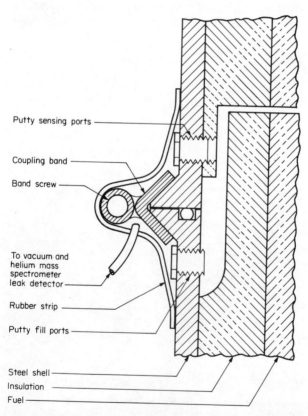

Putty sensing ports

Coupling band

Band screw

To vacuum and
helium mass
spectrometer
leak detector

Rubber strip

Putty fill ports

Steel shell

Insulation

Fuel

Figure 15.2 Proposed *Challenger* joint-and-seal assembly.

band similar to those made for many years by the Marmon Co. Leak testing is done by wrapping with a rubber band, applying vacuum from a standard mass-spectrograph helium leak detector, and releasing some helium inside the rocket chamber. Ports are provided for injecting the heat-insulating putty and sensing that it is there. Simplicity is achieved by a combination of competence and, above all, the *will* to do so.

16

Iteration and Convergence to Final Design

Design requires an alternation between qualitative thinking and quantitative thinking. (This was discussed in Chap. 1.) As this alternation is repeated (iterated), some ideas are discarded, new ideas and magnitudes are generated, qualitative detail is added, quantitative precision is increased, experimental data are used to verify or change both qualitative and quantitative design, and the design converges to its final form.

Why You Should Do It Yourself

Qualitative mechanical design is done with layout drawings, whether made with pencil and paper or by electronic drafting (CAD). It is common for such drawings to start with sketches made by the design engineer and be converted to scale layouts by a drafter. The drafter may have any title from junior drafter to senior designer, depending on how much he or she is expected to add to the contents of the sketches. The design engineer then reviews the layout and asks for changes, and the trouble begins: "Why didn't you ask for that in the first place?" "Is it really necessary to make that change?" "Are you sure it's better?"

These are the symptoms of the drafter's disease I have named "indelible pencil." Unfortunately it is not terminal unless it is so acute that it results in termination. It is related to having a low innovation index.

One of the reasons for your making changes is your ability to criticize an interim design, not just on the basis of easily described reasons but because of your judgment based on experience, insight, and general gut feeling. Try using that as an explanation to the drafter. I don't know about you, but I am simply not smart enough to design to my own satisfaction the first time.

Finally I tried to bypass the drafter sketch conversion fight by doing it myself, and there was a revelation! Now I change what I want to change because I want to change it. I scrap entire layouts and make fresh starts. I have roaring fights over which way is best, but the fights are entirely within my own head. I make decisions, and they are instantly obeyed. If I want to change a tiny detail (I can be very fussy about details), I do so without fear of starting a conflict. *Then,* when I really know what I want, I turn my layout over to a drafter or designer to clean it up, correct drafting errors, add details I probably don't care about, and make final drawings.

All this is not to say that you should not value and use the ideas and judgments of a good drafting designer; you certainly should, and I always do. But the designer gets a crack at the design after I have made my own best start.

You, the Board, and CAD

I have learned to look upon the drawing board or CAD terminal as a laboratory for trying ideas out to scale. I recommend that you do the same. If it is prestigious in your company to "get off the board," the prestige is wrong.

If CAD drafting is used in your company (and it is becoming universal), you must learn to use it yourself, and that can be a time-consuming and tedious task. If you do not and so require an intermediary between you and the drawing, CAD is a liability rather than an asset because you may end up with good documentation of poor products.

When Do You Stop?

One of the universal problems of design is the decision of when to stop improving (iterating) and release the design for manufacturing. Bright engineers are apt to keep generating improvement ideas forever, but you have to stop sometime and build the product or your company goes broke. (I remember, in my first job, being thoroughly bawled out for proposing a great new idea after tool design for the previous design was under way.)

I wish I knew a formula. The conflict pits the cost in money and time versus the incremental benefits of the latest ideas, usually an apples-and-oranges comparison unless you can estimate both in terms of our universal parameter, money.

A compromise

One way to have your cake and eat it too is to freeze the design for a first batch of product and then phase in improvements in batches. Another procedure is to phase in each improvement at an optimum time for that improvement. Of course, the obstacles to introducing changes after production starts include scrapped inventory, scrapped or reworked tooling, time lag in generating new inventory, documentation, customer notification, and testing cost and time.

The problem is a severe test of management judgment, particularly for managers who *like* engineering.

Some Short Essays
in Engineering Philosophy

Perversity Principle: Murphy's Law

The "perversity principle" states that nature is hostile and life is a struggle against a malevolent enemy. From this we get the expression "Just my luck" when something goes bad. Murphy's law ("If something can go wrong, it will") is thought of as an expression of the perversity principle.

The perversity principle is no more true than the Polyanna principle ("Everything will turn out all right"). What is true is that innumerable random things are happening and that only a few of them are favorable. In communication terms, the environment is noisy and very little of the noise is musical. Murphy's law is quite true, but not because something is trying to get you.

Scientific and engineering experiments must be carefully designed to minimize the impact of such random effects.

With respect to your luck, you will find that you have both bad luck and good luck. If you pay attention to the good luck and are grateful for it, your bitterness at the bad luck will be diluted. Elementary statistics tells you that some people have runs of luck, good or bad, just as dice rolling produces limited runs of the same number. We all know Las Vegas stories. (My favorite is about the conservative, stable, and reliable designer in my company who came back from a weekend in Las Vegas driving a Mercedes.)

Entropy in Human Affairs

Related to the random-noise element in life is the concept of entropy in human affairs. In the physical world energy trends to disorganization and randomness; the ultimate form is heat. In human affairs the same sort of thing happens because people are forgetful and have many motives, thoughts, and new inputs, not all consistent within the individual and certainly not within a group of individuals. Therefore programs, after being set in motion, tend to slow down and deteriorate.

If you understand this, then, without anger or frustration, you will steadily follow up, remind, persuade, and do whatever is necessary to keep the train moving along the planned track.

To Spin on a Dime

The associated requirement is the ability to spin on a dime: to discover that your train is on the wrong track and switch it to the right one. To rigid personalities this is a wrenching or even an impossible action. They cannot accept their own error and cannot change a plan. To other personalities, lighter on their feet, recognition of error, without guilt, and changing to a new course are part of the exhilaration of life.

Unforeseen Consequences

Every decision and every action have many consequences. Some are those you intended, and some appear to your surprise and either pleasure or regret. I have seen references to the "law of unforeseen consequences." One of the rationales for conservatism and inactivity is that there may be unforeseen bad consequences and that, as with random events, most consequences will be unfavorable. You can develop and exercise your powers of judgment and prediction, make or modify your decisions based on your prediction of their long-term consequences, and charge ahead. Or you can just lie there and die.

Formal Logic and Defined Categories

I do not want to be repetitious, so I suggest that at this point you re-read the portion of Chap. 2 dealing with logic, arguments, and the bad habit of rigidly categorizing real-world phenomena which really spread out along spectra rather than fitting into sharp-edged boxes. Your thinking should be much broader than formal logic.

Parkinson's Laws

This is a good place to mention, again, Parkinson's laws of human and organizational behavior. They are somewhat cynical, but they show a great deal of insight into the real world. Among his laws are "Work expands to fill the time available for it" and "Organizations grow at a rate independent of the work done by them." Read Refs. 1, 2, and 3. The books are slender and great fun to read.

People Reliability

The human characteristic I learned to value most as a manager is reliability. I would trade a brilliant but irresponsible engineer for a reliable and conscientious technician any day (I have had both). By "reliable" I mean committed to following instructions regardless of obstacles and to coming back with a report if an obstacle which cannot be overcome is reached.

I know that there are both reliable and unreliable people in all walks of life, but I suspect that there is a cultural element which affects the statistics. I suspect that in Japan what I call reliability and responsibility are trained into people from childhood and that this habit is part of the reason the Japanese are trouncing us so badly.

Conceit: The Destroyer

Conceit, or vanity, is the worst single source of failure I have seen. Conceited people think they know more and have better judgment than their supervisors and associates even if they are new to the job and have less education and experience in the kind of work it calls for. They evade direction and ignore advice: they think they know better. They are unsuccessful. Conceit is sometimes accompanied by "machismo," arrogant aggressiveness, and sometimes also by a lack of ethics. The combination is ruinous to a career and, if not eliminated, to a company.

Attributes of the Successful Engineer

There are six attributes which you can cultivate in yourself if you try long and hard. They are:

- Judgment
- Insight
- Persuasiveness
- Ingenuity
- Will
- Prediction

If you get to be good at these, you will be very successful very soon.

Study the Designs around You

I strongly recommend that you continually study the designs of the products around you. Why were they designed as they are? What is wrong with them? How would you improve them? This is pure exercise—the world will not correct itself because you write in a suggestion—but if you are a committed design engineer, you will enjoy doing so and your own skills will be improved all the time. What products? Buildings, cars, appliances, TV, radio, and all their accessories, furniture, lamps, pencils, etc., etc.

If you are interested in organization, do the same with the social institutions and businesses around you: governments, restaurants, social parties, everything. The exercise is good for you, and your insight into the ways of the world will continue to improve.

Work on Yourself

The world is full of self-improvement books, and in a way this is one of them. Yet if you take them seriously and don't just sneer at them, you really will improve and so will your rate of success.

REFERENCES

1. Parkinson, C. Northcote: *Parkinson's Law,* 1st ed., Houghton Mifflin, Boston, 1957.
2. ———: *The Law and the Profits,* 1st ed., Houghton Mifflin, Boston, 1960.
3. ———: *Parkinson: The Law,* 1st ed., Houghton Mifflin, Boston, 1980.

Chapter 18

Prediction as a Design Process

Art of Predicting

Prediction is both an art and a habit. The art is based on your understanding of the laws governing the behavior of the thing predicted. Newton's laws govern the motion of physical bodies and enable you to predict them. The much vaguer laws of psychology enable you to predict the behavior of people, although much less accurately. (If you can figure out the laws governing stock prices, you can become very rich.) Therefore you should study the world to understand its laws as well as you can.

Prophecy, which is prediction, has used the positions of planets, the entrails of sacrificial animals, the random throwing of bones and sticks, the shuffling of cards, and the motions of a Ouija board with the same statistics as consulting a table of random numbers. Learning and applying empirical laws of nature do a lot better.

Prediction is also a habit, and one which it will benefit you to cultivate. Regardless of how well you do or do not understand a particular set of laws, you will be better off on the average if you make predictions and act on them than if you simply ignore the future and react to what happens when it happens.

Products only have problems which were not predicted. Problems which are predicted are solved ahead of time and do not materialize as problems. There are predicted uncertainties, which may materialize as real problems. Usually when you predict an uncertainly, however, you prepare for the problem which might appear.

Predicting What?

What aspects of the world are useful for you, as a design engineer, to predict?

Your competitors are designing products, right now, to compete with your products. If you can predict their designs, you can make the products you are presently designing more appealing than theirs to your customers when both come out. How to predict? You can extrapolate the trends in their designs. You can learn all you can about them (see Chap. 14, "Your Competition").

Your markets are changing, and their tastes, budgets, and sizes are useful subjects for prediction so that you can tailor your designs for future desirability (see Chap. 13, "Your Market").

Predict the effects of age, dirt, corrosion, wear, neglect, and abuse on your product. Your predictions will enable you to improve your design to compensate for these effects and to make it easier for your customer to practice preventive and corrective maintenance.

The technologies which will become available, both for product design and for manufacturing, will influence your current designs if you read the technical news in your journals and plan to take advantage of your predicted availabilities.

When you deal with people, either inside your company or outside, if you predict their responses to what you say or write or do, you will fashion your words and actions to get the most successful results.

In your personal life, you will make investment decisions. Remember that in placing money no decision is, in effect, a decision. The better you predict the future of alternative investment opportunities, the more money you will make (or the less you will lose). For your family and social affairs, I leave the matter as an exercise for the student.

Developing and practicing the art of prediction is a self-discipline but one with a very large payoff.

Specifications, Codes, Standards, Contracts, Laws, and the Law

We are awash in an ocean of general and specific specifications, codes, standards, and laws. Some of them apply to anything you design. I will try to give you an outline of what is out there and some guidance to finding those which apply.

Generic Term: Specs

In this chapter I will use the common expression "spec" to refer to specifications, codes, standards, laws, and contract obligations.

Almost every specification refers to other specifications, which may refer to other specifications, so that you actually face "trees" of specifications, all requiring your compliance.

You will usually be informed of the applicable specs by your managers, but they may miss some. If you do not comply with an applicable spec, you will have to redesign, with cost to your organization and criticism of yourself whether or not it was your fault. It is good practice to assure yourself independently that you know all applicable specs.

Kinds of Specs

Contracts

There are requirements in your company's contracts with customers, vendors, licensers and licensees, and labor organizations which have the force of specifications.

Military specs

A major class of specs consists of military (MIL specs) and other federal specs which apply to federal purchases. Nonfederal customers may recite MIL specs as applying to their purchases as a way of securing high quality despite the perhaps unnecessarily high cost which may come with it.[4, 5]

Trade associations and professional societies

Many specifications and standards are published by trade associations and professional societies. Among the professional societies that issue standards and codes are:

American Society for Testing and Materials (ASTM)

American Society of Mechanical Engineers (ASME)

Institute of Electrical and Electronics Engineers (IEEE)

Among the trade associations are:

American Gear Manufacturers Association (AGMA)

Electronic Industries Association (EIA)

National Electrical Manufacturers Association (NEMA)

National Machine Tool Builders Association (NMTBA)

There are many others; these are named as examples only.

Some of the standards issued by professional societies have the force of law if a legal controversy arises. Failure to conform to specs can have devastating costs if there is a product liability lawsuit against your company and the plaintiff can argue that an injury or a loss occurred as a result of your failure to conform to a code or standard.

Laws

There are many specifications and codes of practice which appear in acts of legislation. For example, there are federal and state antipollution laws and safety laws with special enforcement agencies such as the Occupational Safety and Health Administration (OSHA) and the

Environmental Protection Agency (EPA). Government agencies may be authorized by law to issue their own regulations, which, in effect, are additional laws. There are state, county, and municipal codes. It is quite possible to make a product which is fully legal in the place where it is built and is illegal, in details, in the place where it is to be installed and used. For example, the electrical code of Santa Monica, California, is stricter than the National Electrical Code, which applies almost everywhere else. Use your company's attorney as a consultant in such matters.

Your company's specs

Your own company probably has specs which apply to the product you are designing. Company specs may exist for materials, tolerances, preferred components, manufacturing processes, etc.

Commercial organization specs

Some commercial organizations publish general specs for their own particular reasons. An example is the National Electrical Code, which is issued by the National Fire Protection Association to reduce fire losses and which is the basis of most municipal codes. Another example is the set of specs issued by Underwriters Laboratories (UL), whose business is testing products for safety. (You should find out early whether your product will require UL approval. It may consume a good deal of your design scheduled time and some of your budget to get UL tests and approval.)

Vendor specs

Your vendors issue specs on the properties, limitations, and uses of the materials and components they sell to you. If you violate these specs, you invalidate their warranties and you may make your own product not work or be unreliable.

Customer specs

Your customer may issue a spec on the product it is purchasing, and this in turn may refer to other specs, including general specs for products purchased by that company. The automobile companies, the machine tool companies, and many large corporations have elaborate general specs. Your customer may be your own company if you are designing a new standard product or special equipment for in-house use.

Foreign specs

If your product is to be sold outside the United States, there are many foreign specs to meet. Some are international, and some are national and local. Among organizations issuing specs are the Canadian Standards Association (CSA) and the Deutsches Institut für Normung (DIN), the German standards institute. Metrication is an obvious spec for products going to countries not using English units; there are specs for signs using nonlanguage standard symbols, and there are specs for different voltages (e.g., 220/50).

Patents

In a sense, other people's patents are specs on what you may *not* do. Your company is exposed to lawsuits for "unsuitability of your design for its intended use," for lateness in delivery of the product, and for breach of contract if your design does not conform to all the contract specs. Your company can be sued for patent infringement, so you should get assurance from your manager that an infringement patent search has been made and that no infringement has been found. (In practice, this may really be your suggesting to your manager that such a search be made, and rather early in the program so you have a chance to "invent around" potential infringements.)

Unwritten specs

In addition to the expressed requirements of written specifications, standards, and codes, there are unwritten specs: the requirements of the culture, conventions, and fashions of the market to which you are selling. This is true not only in the field of consumer products but also in technical and capital goods markets. (See Chap. 13, "Your Market.")

Classes of toys come into fashion and go out of fashion as rapidly as fashions in clothing. The same is true of fashions in military hardware, airports, consumer electronics, and numerous other products for which the change in fashion is unrelated to development of technology.

American National Standards Institute (ANSI)

A single source for almost all general specs other than MIL specs is the American National Standards Institute.[2] You should buy its catalog if your company does not already have it. Among the hundreds of specs listed are Drafting Practices Standards, which your company may choose to adopt independently of any specific design project. ANSI will send you, on request, publications describing its standards in detail.

Common Law: Torts

You and your company are subject to the common law. If you are responsible for injury or property loss to someone else (a "tort"), you can be sued for damages. To you as a designer this boils down to whether a tort is due to your faulty design. This is no small matter. Damage awards now can run into millions of dollars. Theoretically you personally are liable as well as your company. The usual form of lawsuit which you should worry about is called a product liability lawsuit. What should you do about it?

Defenses

When designing, keep safety always in mind. Predict all possible modes of failure, and include provisions to protect users and bystanders in case of each of them. Learn all applicable OSHA rules. Keep records of your design studies and tests so that you can testify in court that you have acted properly. ("Negligence" is the ultimate dirty word.) If you design equipment for your own or customers' factories, you are directly subject to OSHA rules and inspections.

Employees

You may interview potential employees, and you will certainly supervise other employees sooner or later. There are laws with severe penalties if you do it wrong. Most of these laws deal with discrimination for sex, race, or religion. You should get trained in what these laws are as they affect you.

Your P.E. License

Most states require that drawings for installations affecting public health or safety be signed by a licensed Professional Engineer (P.E.). It may be in your interest and your company's interest to get your P.E. license.

Please note that the above is only a fragmentary list. I have presented it to show the kinds of things applying to your projects which you had better find out about.

Specifications are not just obstacles to be overcome. Implicit in many specifications is an education on particulars of good design.

Dealing with Customers

Customer specifications can be sources of anxiety and great effort when you try to interpret and conform conscientiously. You can face ambiguous wording and difficulty in securing interpretation from customers themselves. I had this experience with giant corporation J in negotiating the purchase specification for a large and complex robot. I was told by the specifying engineers that the generic specification for machinery, which was ambiguous, was written by other people whom they did not know, that it would be enforced at acceptance time by still other people whom they did not know, and that they themselves were powerless to waive or interpret these provisions for me. We did our best, took some chances, and fortunately were successful.

Customer enforcement varies from lax to maliciously diligent. I was told by machine tool company I, the system contractor that was the direct purchaser of the above robot for incorporation into a machining cell for J, that it had once been compelled by J to rewire a group of lathes, already installed and working well, because the wire colors in the lathes had been discovered not to match the wire colors in a referenced specification.

I suspect that strictness of enforcement varies with how urgently the customer either wants the machine or wants to get out of the deal.

(You may feel that such trivia are beneath your dignity as a professional engineer, but your company comptroller will be glad to explain to you why he or she disagrees and why getting paid by the customer is so important.)

Buyers' specs describe what they think they want and what they think they can get for what they think it will cost. They may be wrong on many counts, as revealed by the design process. Certain requirements may turn out to be much more difficult to achieve than expected and others unexpectedly easy to exceed. As you work on the design (or proposal), some desirable but unspecified features may be invented by you and others may become, in your mind, not very useful, at least as specified.

The relationship which gives customers the best and the most for their money and gives the vendor a reasonable chance for a reasonable profit is one in which spec and price are in continuous renegotiation from the beginning until the product is accepted. Sometimes a proposed spec change may give the buyer more or better product and the seller more profit at the same price; sometimes it may give the buyer more or better at an acceptable increase in price.

In the real world there is a conflict between buyer and seller about how much product the buyer gets and how much money the seller gets. It is up to your marketing department (which may be you if you are an entrepreneur) to know the various procedures of bid to price, proposal and bid to general spec, unsolicited proposal (usually after informal proposals and discussion), proposed renegotiations of spec and price, and the many incentive formulas of government contracting.

You may find an opportunity to practice your persuasive power on your marketers to get them to take a renegotiating initiative when you have a great new idea.

As I describe elsewhere, when I was a space vehicle proposer, I responded to a NASA RFQ for a geodetic satellite with both an exact response to the specification and an unsolicited proposal for what I felt was a better design (Fig. 13.1). The customer completely agreed with me. It canceled the RFQ and copied my drawings into a new RFQ, and a different company won the contract on price. The customer's ethics are one of the risks of the marketplace.

Understanding customers

Let me illustrate the importance of insight into the customer's *real* needs. When I was still working in the Convair analog computer laboratory, my boss, an enterprising man and a wonderfully likable supervisor, brought in an RFQ from the Air Force for an analog function generator for multiple functions of two independent variables. The spec was obscure as to meaning and intent, but we put 2 and 2 together and inferred that the Air Force probably wanted the generator to generate many empirical functions of speed and altitude for simulating the performance of aircraft. I invented a system to do just that, with a very convenient way of entering the data at a series of points and an interpolator among those points.

Our group had never proposed on a contract before (we provided an in-house service), so we had no drafters, no standards, and no conventions. We divided up long roll-size sheets of paper into random-sized outlines for individual drawings, did some drafting freehand, and in general would have driven any respectable chief drafter to hysterics.

We later met with the customer engineers. They said, "We got proposals from all the major companies in the business, but none of them understood what we needed, and none of their proposals were useful. We were feeling depressed. We didn't even want to unroll your big sheets, but the law required it; so we did. They looked terrible, but we started to study them as we were required to do. We kept nodding our heads as we went along because you clearly understood what we really needed and gave it to us. There was no choice; we had to award you the contract." [We built two generations of FOXY (*functions of x and y.*) The first was electro-mechanical, and the second was all electronic.]

Negotiating

Do not assume that you should only look after your own interest in negotiations and that it's the customers' responsibility to look after theirs. If they have not done a good job for themselves, sooner or later their people will get angry and blame you, and there will be a big fight before you get paid.

For example, this happened to me: In the early days of my robot company I contracted with major toy company N for an assembly machine. Speed was not specified in the contract. We built the machine, and it worked. The company's acceptance test engineers came to our plant, were pleased with the operation, pulled out their stopwatch, and panicked. Unless the machine went faster than it could go, the company could not meet production goals. (A postmortem of the history showed that the company's chief engineer was suffering from some kind of mental illness, had put in a speed requirement, and then, without explanation, removed it before we ever saw it.) I visited the company's plant with movies of the machine, appealed to its president (who turned out to be an honorable man), and got paid. The machine was scrapped. I blame myself almost as much as the toy company for not raising the question of speed in the first place.

Customer's Duties to You

There is a reverse aspect to the meeting of specs. Specs also require customers to provide parts, test equipment, environment, and interfaces with which your equipment is designed to operate. Insist that customers meet their part of the spec, or you will be undertaking the impossible.

For example, I learned this the hard way, too: At Numerical Control we contracted with dental company O to make an automatic-assembly machine for a device containing three plastic parts, two metal parts, and a measure of mercury. The company provided parts to use in machine development, including plastic parts made in temporary molds. The machine worked fine (Fig. 19.1).

The company then sent us plastic parts made in permanent molds for final test, and my engineer said he was having some problems. He was very bright and I was busy, so I left him to do his job without interference. I put off the customer and put him off until he became furious. Finally I decided that enough delegation was enough and looked at the job. Some of the "good" parts from the permanent molds were distorted and had flash, so they jammed in the machine. The engineer felt that he had been given the parts he had to assemble and he would assemble them if it took forever and killed him in trying. I called a screeching halt and notified the customer. By now the company was so mad it didn't believe me and wouldn't even look. It pulled out the job.

One last example story: We contracted to build a pair of large, complex, and advanced robots for a machining cell being made by major

Figure 19.1 Automatic-assembly machine.

machine tool builder P. At the same time the chief engineer of P retired and was replaced by his assistant. As P got into more and more trouble at its end, it "interpreted" the spec to add more and more equipment and responsibility to our end. Its vice president backed up the chief engineer, frankly admitting that he did so only to save money. The new chief engineer was a very poor engineer, was an inexhaustible excuse maker, and had a staff of incompetents whom I referred to as Abbott and Costello. I had hoped to make P like us because it was a potential source of a great deal of business. Instead I relearned the lessons of appeasement. (Please refer to the history of Chamberlain and Hitler.)

Please don't be overwhelmed by the world of specs; just take them as seriously as you take engineering analysis.

REFERENCES

1. Berke, Lanny R.: "Protecting against Product Liability Suits," *Machine Design*, Sept. 24, 1987, p. 194.
2. *Catalog of American National Standards*, Sales Department, American National Standards Institute, 1430 Broadway, New York, N.Y. 10018. Telephone: 212-354-3300.
3. Davis, Philip M.: "The Designer's Responsibilities," regarding product liability, *Design News*, Feb. 23, 1987, p. 142.
4. Department of Defense: *Index of Specifications and Standards (DOD-ISS)*, Subscrip-

tion Service, Superintendent of Documents, Government Printing Office, Washington, D.C. 20402-9325.
5. Navy Publications and Forms Center, Philadelphia, Pa. To purchase MIL specs, telephone 215-697-3321.
6. Struglia, Erasmus J.: *Standards and Specifications Information Sources,* Gale Research Company, Book Tower, Detroit, Mich. 48226, 1965. Out of print.
7. Trachtman, M.: *What Every Executive Better Know about the Law,* 1st ed., Simon & Schuster, New York, 1987.

The Pervasive
Parameter: Money

Psychology of Money

Psychiatrists will tell you that money, power, sex, and religion are the strongest human motives. To a considerable degree money is a form of power, and it is "the love of power [which] is the root of all evil." The highly emotional word "greed" means a thirst for money. I have listened to venture capitalists speak admiringly of the greed of their clients, thereby expressing their own greed. I have frequently watched smiling, happy faces turn suddenly grim when a question of money going to or away from them arises. Highly intelligent, educated, and logical minds suddenly issue irrational nonsense to prove that money should flow in and not out.

Cost as an Engineering Parameter

I had a professor in engineering school who taught that "an engineer is a man who can do for one dollar what any darn fool can do for two."

Cost, in engineering language, is a figure of merit. High cost is bad; low cost is good. Cost is a common denominator of value and is usually expressed as money, which is the common medium of exchange. Cost is what apples and oranges have in common.

At one time the term "cost-effectiveness" was used. It is more appropriate for engineers to think of "cost efficiency," defined as the ratio of the value of the product out (usually not measured in dollars) to the cost of the effort in (measured in dollars).

We should face reality and learn to be cost-efficient in our work; we should study and treat cost as an engineering quantity as we study energy and voltage and force.

Many engineers think of themselves as members of a profession purer than the occupation of a businessperson. I must tell you that this is naive; all engineers are businesspersons whether they like it or not, just as they are politicians whether they like it or not. The basic business transaction of design engineers is to sell their services to their employers. The ruling principle of this ongoing transaction is that the employers pay money to the engineers to motivate them to make somewhat more money for the employers. Otherwise, why should the employers continue to pay?

Every engineer who wishes to earn a higher salary, every academic who wishes to achieve tenure and the financial security it brings, every inventor who hopes to get rich on royalties, every lawyer and doctor who accept fees, every not-for-profit organization which seeks to grow and generate power, security, and pay raises for its managers, all are motivated, at least in part, by the same desire for money. It is the same motive that drives the merchant, the manufacturer, and the speculator. You may rank your own route to this goal as morally higher than another's because money gain is not your only motivation and perhaps not your principal one; I would agree with you and share your pride. But we are all businesspersons too, and we are paid to share the business motivation of our companies. That's our fundamental job.

Components of Product Costs

Market research and proposals

The initial cost of a product is in its market research and, sometimes, in proposal engineering and sales. You may or may not participate. These costs result in the decision to make the product.

Engineering

The cost of engineering is for your time, that of your managers and your assistants, and that of the computer and experimental work performed to aid you in producing a design. Engineering cost also includes the ongoing cost of your time to make modifications and solve problems as the product enters and continues in production. For the same design value, the lower the engineering cost you generate, the better the job you have done, the more money you company will make, and the more you will be appreciated and rewarded.

Manufacturing

The manufacturing cost of your product comprises labor, material, and the cost of capital for inventory and capital equipment. As a design engineer you have great power to produce a better product for less manufacturing cost, i.e., to make a more cost-efficient design.

Part of the manufacturing cost, which I intentionally separate, is the tooling cost. It, with your engineering cost, will be amortized over the total production run. In your mind, as you design, you must continually trade off higher tooling cost and lower part cost. For example, a die casting has higher tooling cost and lower machining cost than a hog-out, which has lower tooling cost and higher part cost. Another part of manufacturing cost, which I intentionally separate, is quality control, i.e., inspection. In your design thinking, bear in mind what inspection processes will be required; they may influence design decisions. Also think of the gauges and test equipment which should be used. You may be able to make a substantial contribution by suggesting and designing such inspection tools.

In Chap. 2 we discussed the continuing challenge to conceive two-way winners in which, in this case, there is both low tooling cost and low part cost. For example, a cleverly designed stamping might replace both the die casting and the hog-out.

Marketing cost is something that engineers do not like to think about but that general managers must consider as equally important to engineering and manufacturing costs. It includes the initial market research and proposal engineering and sales mentioned above, plus the costs of salespersons, advertising, packaging, warehousing, and warranty. If you are conscious of these costs, you may be able to modify your product design to reduce them. If you think like a marketer, you will make the product so its price can be low without sacrificing value and thereby encourage customers to buy. If you make your product more reliable, the warranty costs will be less and you will further encourage customers to buy. If you design the product to be easily and safely packaged against damage, the cost of replacing damaged goods will be less.

Overhead

Overhead is an irritating abstraction to engineers, but it is a real cost to your company over which you have some influence. Overhead includes the cost of rent, insurance, and internal services by such people as buyers, accountants, and managers. There is not very much you as a designer can do to minimize these costs, but occasionally there is an opportunity. Depending on your design choices, your purchasing department spends more or less time in buying the components and materials for your product. You will be saving your company overhead costs if you select readily obtainable standard components and materials with multiple reliable vendors and if your design minimizes the total number of vendors with whom your company must deal. To use one of the words which frequently recur in this book, it is a call on your judgment to make only minor sacrifices in the quality of your design to simplify the purchasing process.

Capital

Your company spends money (invests capital) on the product continually from the moment when it is a gleam in the eye until there is enough profit on sales to equal the capital invested plus the cost of that capital. It sounds strange to say that money costs money, but it does. If the money is borrowed, the cost is interest. If it is stockholders' money (or your own in your entrepreneurial business), it could otherwise have been loaned out at interest and the interest you are not getting amounts to a cost. Therefore anything you can do in product design or project management to spend less time and need less machinery, tooling, and inventory directly helps the company's profit on the product. With a little luck and self-advertising your success will let you share the benefit.

Transportation

The cost of transporting materials and components from their vendors to your factory is paid for by your company and is part of the cost of manufacturing. In your design decisions you have some control over weight, transportation mode and distance, and fragility of the incoming material and components and in so doing have some control over this portion of the cost of the product. Outgoing transportation is paid by your customers, but it is viewed by them as part of the price of your product. Your design has some influence on outgoing transportation cost. Total weight, fragility as packaged, and separate packaging of dangerous or sensitive materials all affect cost. For example, in my robot company, in which we made very large machines having dimensions up to 150 ft, we designed the machines to be easily disassembled and reassembled. They were then packed, without crating, in air-ride moving vans which carried them with very little shock and vibration and no additional handling from our factory door to our customers' factory door. The result was a combination of no crating cost, no shipping damage, and short shipping time but high price per ton-mile. *This shipping system was an integral part of the design.* Both our customers and ourselves were quite pleased with the results.

Packaging

Bear in mind, in your design, that the packaging of your product, whether a decorative perfume box or an oceangoing crate, is part of the design of the product and contributes to the cost of the product. (See Chap. 32, "Design for Packaging and Shipping.")

One-Time versus Per-Unit Costs

Costs can be divided into one-time cost and per-unit cost. The accounting words used are "nonrecurring" and "recurring." For example, most engineering is a nonrecurring, or one-time, cost; most material is a recurring, or per-unit, cost. In the case of a product which is made as a single copy, such as a manufacturing machine limited to your own factory's use, all costs are nonrecurring and there is no distinction. Your economic objective as a designer is to minimize total cost through the anticipated production run.

Your Customer's View of Cost

It costs your customers to operate your product. They will consider operating cost as part of the life-cycle cost when deciding whether or not to buy your product. A machine which requires a skilled operator costs more to operate than a machine which requires only an unskilled operator. A machine which requires two operators costs more than a machine which requires one. A machine which requires frequent replacement of expensive components and supplies, such as special fuels and lubricants, costs more to operate than a machine which requires few replacements and supplies, both of which are inexpensive and readily obtainable. A machine which is more energy-efficient costs less to operate than one which is less energy-efficient.

To your customers the performance of the product can be measured as a cost. A fast machine is more cost-efficient than a slow machine. An accurate machine has cost benefits as compared with an inaccurate machine.

The cost of maintaining your product is of great importance to your customers and is part of their life-cycle cost. This cost includes both preventive (routine) maintenance and corrective maintenance (in case of a failure). Maintenance cost is affected by the skill required, the special tools required, the cost of replacement parts and supplies, and the time out of service required for maintenance. (Please refer to Chap. 22 for a discussion of kinds of reliability and their effects on maintenance.)

Life-cycle cost

A valuable way for your customers to consider the cost of your product is its life-cycle cost. This is the original cost plus the operating cost plus the expected maintenance cost for the anticipated life of your product. From the customers' point of view it may be more desirable to spend more up front and have less operating and maintenance cost, or, if their available money is limited at present, they may prefer a lower initial cost at the expense of higher operating and maintenance cost. Furthermore, if they anticipate major modifications of the product as time goes on, they may well prefer a lower initial cost since there will be less cost thrown away when modifications are made. The importance of all this to you, the designer, is that you should get in the habit of thinking like your customers so that you will please them with your product to persuade them to buy it, endorse it to other prospective customers, and come back for more.

Some Ways to Reduce Costs

It is your responsibility to design for low cost; in almost every case your employers want you to do so and in their mind are paying you to do so. We sometimes hear of special market situations in which high cost is deliberately sought, but these are most unusual.

It's not easy. Anyone can make a product a little cheaper and a little worse. (Anyone can also make it a little better but a little more expensive.) Your challenge is to make it just as good or better and a little cheaper too. This is an ongoing struggle for two-way winners. There is as much professional challenge in cost reduction as there is in making the product work in the first place.

Here are some techniques you can use:

Better tolerancing

Study the dimensional combinations and tolerances of your parts. Can your assemblies be designed so that looser-tolerance parts will work? Are the tolerances stated on your drawings really the tolerances needed, or are they arbitrary "standard" tolerances which add cost without having real value?

There is a great tendency on the part of drafters and junior designers to use standard tolerances. Standard tolerances are usually either too tight or too loose. You may not want to spend the time to do a complete tolerance analysis for every detail drawing. However, you should scan all drawings looking for tolerances which may add unnecessary cost. When looking for tolerances, also look for dimension arrays (which point is dimensioned from which reference point) to permit a system of dimensions and tolerances having the least cost to manufacture. In making a tolerance study bear in mind the nature of the process to be used in making the part. Parts made in molds and dies have uniform dimensions, but those dimensions will all have tolerances.

Other processes, particularly those involving human operations in setting a dimension, are subject to human skill and must be tightly controlled by tolerances and inspection. Inspection itself is a cost, and reducing this cost and the cost of inspection errors is one of the motives for automation.

There is often a great deal of benefit to using match notes on the drawings of parts which go together instead of tight tolerances for the individual parts. You may want to line up two holes in part A with two holes in part B within 0.001 in, but you don't care whether the space between the holes varies from its nominal as much as 0.100 in. Attending to such considerations can save very substantial amounts of money in the cost of your product with no loss of quality.

Minimize part count

In electronic circuits minimizing the part count not only reduces cost but increases reliability. You can do it if you really try. In mechanical devices you can often combine parts into a single part. (There is more on this subject in Chap. 15, "Simplicity," and Chap. 24, "Improving Existing Designs.")

There is a design doctrine which says that parts should be dimensioned for function without regard for manufacturing processes. This is a costly oversimplification. You should think of both and design for both.

Value engineering

Some years ago someone coined the term "value engineering" and built a career on it. Value engineering is simply second-guessing a design to make it better and cheaper. It is a good thing to do, although one should remember that it is itself an additional engineering cost. You should do it all the time while you are designing. (Personally, I have a distaste for coining a new word for an old thing, pretending that you have invented a new thing.)

Design for easy assembly and automatic assembly

Have you designed for easy assembly? Have you visualized the assembly procedure as if you were an assembler on the bench? Have you designed for automated assembly if your quantities suggest the possibility of such manufacturing technology?

Which part-fabricating technology?

Have you studied all possible techniques and materials for making each part or combination of parts and the design variations which permit the use of each technique? (For example, you can design the same part to be machined, stamped, forged, die-cast, lost-wax-cast, powder-metallurgy-formed, etc.) Do you really know all the manufacturing techniques and materials which apply to your products? Have you studied vendors whose specialized techniques can reduce your costs? Learning new and unconventional manufacturing techniques and specialized vendors should be a part of your continuing self-education.

Simplify

Is the design really as simple as you can get it without loss of function? Complexity is easy; simplicity is the result of great effort. This is a universal principle in science, engineering, mathematics, art, legislation, and the design of business deals. Chapter 15 is devoted to simplicity.

Law of Diminishing Returns

The law of diminishing returns says that in almost all activities a point is reached in which further input does not produce a proportional output. This applies to the training of athletes and the engineering of cost reductions. How do you know when to stop? Sorry, but I must cop out and refer you to our old friends experience and judgment.

Designing for the Right Price Market

For many products there are high-price markets and low-price markets. You and your marketing organization may choose the portion or portions of the price spectrum for which you will design. It is common to design a line of products with many common parts and features with more or less superficial changes to convert them from one part of the spectrum to another. Note that your marketing department's price strategy may have little resemblance to your manufacturing costs.

For example, Jack Rabinow, to whom this book is dedicated, invented the straight-line-motion phonograph arm and founded a company, Rabco, Inc., to manufacture and sell it. He chose the low-price mass-market portion of the spectrum as his marketing target. In order to make large quantities at low prices in a new factory he encountered start-up cost problems which he would not have had he gone to the high-price end of the market. His conclusion was that he should have made small quantities of very elegant design and sold them at high prices. They would willingly have been paid by a small but eager and affluent market. He would then have been profitable, would have established the prestige of his product, and would have been able to exploit the low-price end of the spectrum. Hindsight is wonderful.

The longer you live, the more ways you will learn for getting the most product for the least cost; there is no end to the study of cost reduction. The important message of this chapter is that you should be *interested* in cost reduction and *want* to do it.

21

Quantity Effects
on Design

Basic Principles

The way you design a product should depend on the quantities of the product to be made. If you are to build one special machine or electronic system, it is less costly to use established engineering practice to make a product which will work right the first time, rather than to undertake extensive R&D, and to design for manufacturing processes which do not require large tooling investments (e.g., molds, special integrated circuits) even though it would be costly in quantity production. On the other hand, if you are designing a mass-production consumer product, it pays to do a great deal of R&D and to design for very low per-unit cost at the expense of large tooling investment. In accounting language, the total nonrecurring cost plus the total recurring cost for the complete production should be a minimum.

Business Considerations

This simple principle is modified in the real world by several other considerations which are matters of judgment rather than of calculation. Is there enough working capital available to pay for a large initial tooling investment? How sure is your company that it will really sell a large quantity? What is the probability that you will want to or have to change the design when you are only partway through the predicted production quantity?

Models

A perpetual problem is how to make single models and preproduction short runs of mass-production products without the full investment of both time and money required for production tooling. In electronic circuitry a special integrated circuit can be reliably replaced by standard components during R&D at the cost only of physical space. In mechanical manufacturing it is much harder without tooling to make parts and large assemblies which are equivalent to tool-made parts.

Many processes have been developed to solve portions of this problem, and it is worth the time to learn about them. I will tell you about some that I know of as an introduction to the subject.

Eleven short-run techniques

1. Single-cavity "soft" tooling can be made for plastic molding and die casting at a fraction of the cost of production tooling. The molds may require much manual work to assemble and disassemble, and the part may require some hand finishing, but the total cost is relatively low.

2. There are companies which specialize in short-run stampings. The dies employ no die sets, and their use requires much handwork per part, but they are surprisingly inexpensive.

3. Die-cast parts can be simulated by sand-cast or lost-wax-cast parts with machined finishing. Molded plastic parts can be simulated with cast plastic with machined finishing. Of course, almost any part can be "hogged out" of a solid block by machining. For small to medium quantities, NC machining may be surprisingly economical. There is, of course, the risk that for highly stressed parts the strength simulation will be poor, either too bad or too good, and material substitution may prevent correct simulation.

4. Small quantities of lost-wax castings ("precision castings") can be made from expendable patterns machined from polystyrene components cemented into complex patterns.

5. Flame-cutting and welding plate stock can simulate large iron castings as well as being a valuable production technique.

6. Complex solids can be made by printing and etching thin sheet metal and soldering or cementing laminations into a solid.

7. Simple parts such as lengths of tubing can be cemented, soldered, brazed, or welded into simulations of complex parts.

8. There is a new technique in which complex plastic parts are

made by progressively building up layers of liquid plastic hardened by a point of ultraviolet light which scans a container of the liquid under computer control.

9. Remember the availability of some unconventional machining systems, usually employed for toolmaking, such as electrical-discharge machining (EDM) and laser machining.

10. Kits are sold for making experimental printed circuits in your own laboratory in a very short time.

11. There are breadboarding boards and terminals for simulating printed circuits.

Designing for short runs is, in many ways, a greater challenge to your ingenuity than designing for mass production.

Reliability and Maintenance

Reliability is a good thing, like motherhood, patriotism, and the flag. However, there are different kinds of reliability needs.

Different Service Requirements for Reliability

In a newspaper printing plant it is permissible that the machinery require careful, skilled, unhurried, preventive maintenance every single day; *but* for a few hours each day after the press has been turned on it is absolutely essential that no breakdown whatever takes place because the cost of interrupting a newspaper printing run is enormous. A surgical heart-lung machine has similar design rules, but here a failure can cause a death.

A cardiac pacemaker can be maintained by surgical replacement, but if the failure is sudden, it can threaten a life. If the failure is "fail-soft" and gives warning, there is time for surgery before there is a life threat. Battery wear-out is managed by surgical replacement of the entire pacemaker on a time schedule. Therefore the longest-life batteries available are used, and the batteries as well as the remainder of the pacemaker are made under extremely strict quality control.

A different kind of thinking is required in the design of space vehicle hardware, for which maintenance by humans is zero, a small amount can be done by remote control, operation is continual, and the cost of failure of an inexpensive part can be the multimillion-dollar cost of a lost mission.

Domestic appliances must work with almost no preventive maintenance, but an occasional random failure is acceptable.

Kinds of Maintenance

Consider the availability of skill, tools, and parts inventory and the consequences of random failures. Is your company policy to correct failures by part replacement, by subassembly module replacement, or by complete disposability and complete replacement? Are modules repaired by the factory and returned to a replacement pool? What replacement parts, modules, and supplies should be stocked by the customer?

In your product's maintenance environment will the people be trained? Untrained? Equipped with an inventory of spare parts? Equipped with good maintenance tools? Disciplined and goodwilled or undisciplined and malicious? Trained maintenance people employed by your own organization or by distributors for your organization? In this case you can assume that the maintainer is unlikely to do damage, understands the machine, is equipped with special tools, and is capable of rather educated diagnostic thinking. Next there is the class of professional maintainers such as garage mechanics who are more or less trained and more or less well equipped. Third, there is maintenance by the user, who may be highly qualified, as in the case of electronic technicians in research laboratories, or may be highly unqualified, as in the case of users who include the uneducated and unskilled.

Disposability

The use of disposability as a maintenance technique has been increasing for some time. In medical products disposables eliminate the need for sterilization between uses and the handling and reassembly of components into kits. The labor saving in hospitals is substantial. There is a continuing development program in this industry to make instruments in very inexpensive ways to justify their disposal. The single-use hypodermic needle is the most commonly employed example. In both consumer and other maintenance, the use of disposables reduces the skill and cost of maintenance. Your design policy in this matter deserves great consideration.

Accessibility for Maintenance

Accessibility for maintenance is of great value in reducing the cost of maintenance and reducing the irritation of your customers when they have to go through contortions and squintings in order to repair or adjust your product. Accessibility for diagnosis, adjustment, lubrication, and replacement prolongs the life of the product because it makes it easier to maintain and people are thus more likely to do so. It also promotes goodwill from the customer who has to perform the maintenance. However, such maintenance accessibility can conflict with aesthetics, the size of the envelope, and the cost. Here is another opportunity for two-way winners.

In my robot company I insisted that all components subject to possible replacement be surface-mounted instead of being buried in machine cavities. Such burial is quite common, primarily to improve the appearance of machines. We had the further advantage that surface mounting permitted surface wiring and plumbing with consequent lower initial cost and further ease of maintenance. For industrial service this is rarely a disadvantage. However, there is always an exception; this was ours:

We were visited by manufacturing engineers from major cosmetic company R who were looking for an unpacking machine for cosmetic containers. It was obvious to them that our robot would do the job well and cost-efficiently; they were thoroughly pleased with all aspects of the machine.

"However," they said, "there is a little problem." Their company sells cosmetics through part-time saleswomen. To stimulate enthusiasm they invite them to the factory. The saleswomen are picked up at the airport in the company pink Cadillac and escorted through the factory, where all machinery is painted pink. Most of the machines are standard commercial machines which have been developed with a smooth appearance for sales appeal.

The engineers said that, frankly, our robot was the best machine they could find for the purpose but that it was not pretty and would not fit in with the aesthetics of their factory even if painted pink, which I offered to do. They returned to the factory, gave their report, and reluctantly telephoned me to say that they could not buy our robot because it was not pretty enough. You can't win them all.

As the designer you have the choice of designing for special maintenance tools, which makes your design life easier, or working very hard to make maintenance easy.

Built-in Diagnostics

In some modern electronically controlled products it is the practice to design in very complex diagnostics which display to the maintenance person the nature of the fault or the nature of the action required. It is sometimes very desirable to put conspicuous displays of need for supplies such as paper in a duplicating machine so that the user is not tempted to fix a nonexistent fault and thereby create a fault.

Hot-Line Maintenance

The establishment and availability of hot-line operating and maintenance system service permits a degree of complexity in operation and maintenance which might not otherwise be permissible.

How to Improve Reliability

A product may be modified in many ways to increase reliability and reduce maintenance. Better components, exclusion of dirt, simplified adjustments, easier access, redesign of parts subject to failure, derating of both electrical and mechanical components, redundancy of components, reduced operating temperature from better heat-transfer means—the list is endless.

Testing

Testing is essential to expose sources of unreliability. However, you cannot test reliability into a product either in development or in production. Increased reliability comes only from improved design and better manufacturing.

You test to see if the product works at all, to see its performance under different combinations of manufacturing tolerances, to see its performance under different combinations of environmental stress, to determine its operating life, and to see its performance with respect to safety. You test during R&D, and you do partial testing in production to detect defects which may be either initial or "infant mortality."

A discussion of the cost aspects of maintenance is given in Chap. 20.

Planning for Maintenance

As part of your design process you should predict all possible failure modes and decide what to do about each. Aside from designing out the possibility of failure in each mode, you should provide for preventive and/or corrective maintenance for each. For example, a part subject to failure should not be welded or riveted in place.

In electronic circuits part count should be minimized and temperature should be minimized.

You should keep records of your reliability design studies and tests, both for future use and for defense in case of product liability suits.

People Problems

We have considered the reliability of your product and many of the things you can do to make it more reliable. Part of the problem is the reliability of the people, organizations, and procedures involved from initial design through testing, manufacturing, shipment, and use.

The military have elaborate specifications on inspection, shipping, and many other things which affect the final reliability of their purchases. The Food and Drug Administration (FDA) has similar and even more stringent requirements on products it controls. (One of my clients was almost shut down for deviating from these requirements.)

You should assess the reliability of the people, procedures, and organizations that will be involved with your product and request your management to make changes in anything which you fear may reduce its reliability.

Designing for reliability and maintenance is not as much fun as designing for function, but it is an equally important part of designing for success.

Models and Experiments

There are limits to our ability to design successfully by thinking, researching in libraries, consulting, drawing, and calculating. We make models and perform experiments to extend our capability to design. Some of these are as follows:

Eleven Kinds of Models

1. Mental models and mental experiments are thinking techniques for applying your knowledge to a problem by imagining how a physical model or experiment would behave. They are particularly useful in discussion to explain or persuade someone else what would happen in a real situation.

2. Visual analogs look like the product, in either two or three dimensions, and either roughly or accurately, depending on your purpose. They may be made of cardboard and tape, or clay or plastic, or any other material. They are used to visualize the product (either by yourself or by those you want to show it to), and they are used to arrange parts in a configuration such as an electronic assembly, an automobile engine compartment, or an entire factory floor.

3. Some CAD systems present visual analogs on a terminal screen. If you have such a system and are good at using it, it is fast, versatile, inexpensive to use, and easy to change. However, if you depend on a CAD-trained drafter to communicate from you to the computer, you throw away the use of the drafting board as a laboratory in which you

think and change while working to scale. It is tragic to use a computer to make beautiful drawings of poorly thought-out designs.

4. Kinematic "paper dolls" are cardboard cutouts with thumbtack pivots to demonstrate the motion of a linkage. Again, some CAD systems will do this better and cheaper if you have an adequate system and are skilled in using it. However, unless you use a computer application package often enough to justify its original cost, plus the cost of learning to use it, plus the cost of relearning what you have forgotten between occasional usages, it may be better to use an old-fashioned way.

5. Templates are a kind of model used in drafting, usually a plastic sheet with a cutout of the desired form. There are many commercially available templates, from geometric shapes to hydraulic valves. You can cut out your own for a special shape you use often. CAD packages provide the equivalent.

6. Children's construction toys, both static and with moving parts, are used for visualization models. I used to watch engineer B moving a set of vectors, made from a child's construction toy, through the air as he planned the maneuvering of satellites.

7. Rough working models of wood or plastic made many times scale are useful in visualizing the construction and action of miniature mechanisms. Conversely, models to very small scale can help visualize the construction and operation of very large structures and machines. Architects make cardboard and plastic scale models of buildings and terrain for visualization by both themselves and their clients.

8. Test-of-principle (TOP) models are used when you are not sure how a device will work and when you feel it would be too costly or difficult to model it adequately on a computer. Devices sensitive to friction are an example. One can use as much ingenuity in designing a cheap and dirty but adequate TOP model as in designing the product. A TOP model does not even have to look like the real thing if, in your understanding, it tests the principle you need tested. Electrical breadboards are a form of TOP model.

9. Related to the TOP model is the experimental testing device. It may be used only once, it need not withstand service conditions outside your laboratory, it should be inexpensive, and it offers another opportunity for your ingenuity. I would like to describe two examples:

In Convair I designed a transducer to record the motion of the booster section of the Atlas missile as it slid down its rails after its burn. The motion was 10 ft at an acceleration of 2 G. Need for the device was an afterthought resulting from flight test problems, it was needed in a hurry, and it had almost no budget. That sort of thing does not happen in your company, of course. The transducer (Fig. 23.1a) used 10 ft of piano wire wrapped around a drum mounted on a poten-

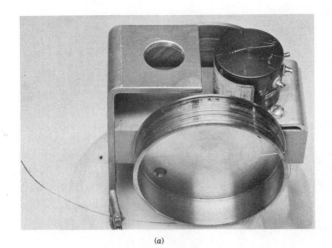

(a)

(b)

Figure 23.1   10-ft transducer. (a) Transducer. (b) Test setup. (*Courtesy of General Dynamics Corporation.*)

tiometer shaft. A spring-loaded drag brake prevented overshoot. How could we test it on the ground by generating a 10-ft, 2-G input with no money?

I hung a sandbag on a cord wrapped around a 2-in-diameter pulley (Fig. 23.1b). The piano wire was wrapped around a 4-in pulley on the same shaft. A cord restrained the pulleys from turning until I had burned the cord with a match to give a shockless start. A pen recorder plotted potentiometer output versus time. The whole rig was set up on the computer laboratory fire escape. It worked once, it worked fine, and the transducer was accepted and used.

The second example was a problem in very precise measurement with low budget and no appropriate instruments available. I had invented a gyroscope transducer which had to measure a few seconds of arc. Figure 23.2 shows the test instrument. I machined a neck in an aluminum bar 1.000 in in diameter and 1.000 in long. The torsional

(a)

(b)

Figure 23.2   Small-angle instrument. (a) Test head. (b) Torque system. (*Courtesy of Rabinow Engineering Co.*)

stiffness of a round bar can be computed easily and without approximation. Torque was applied to the bar with some cheap light chain, pulleys, and a fish scale good to about 10 percent. I could generate a few seconds of arc to about 10 percent of a few seconds. Confirmation came when we measured the scale factor of the transducer both in the few seconds' range and in the few degrees' range where we could use a simple dividing head. The scale factor was the same. The transducer was patented and manufactured by our client.

10. Working models are preproduction real things with various degrees of approximation. They are used for testing, both for detail measurements of stresses and other engineering properties and for overall testing to see if they do what they should and to make experimental changes.

11. A special case of the working model is the one-shot special machine. If it is to be delivered to a customer, the experimental changes must not leave the machine looking as if it had been reworked; it must look as if it were one of a production lot. Sometimes it is worth building a full working model, doing the experimental work on it, and building a new model to ship out. This is a major cost in the special-machine business. On the other hand, if the machine is for your own company's factory or laboratory, the company people may be willing to take it with toolmarks, extra holes, weld burns, and rough wiring as long as it works well.

How to Debug

I have seen so much debugging done badly that I would like to make a few suggestions.

Don't tinker. Debugging requires careful thought and disciplined action, or you will make a mess, waste time and money, mask symptoms, and fail to fix the product.

When a fault appears, *don't clear it and start again!* If you can be assured of tomorrow's sunrise, you can be assured that that fault will repeat, perhaps after the customer gets the product. Look at the fault, examine it, measure it, photograph it. Form a hypothesis about what caused it, and test the hypothesis. (You studied science in college.) Use the art of instrumentation and measurement that you learned at school. I have used every instrument I know from a 6-in scale to high-speed photography. (Did *it* produce a revelation!) If you do not have the needed instrument, you can rent it or hire a consulting service to make the measurements.

I would like to suggest that your fingers and hands are excellent seismographs to sense small vibrations and shocks in moving-part products. Furthermore, your brain can relate the timing of what you feel, see, and hear and thereby lead you very quickly to diagnoses.

If the failure was transient, get ready with the necessary recorders to register it on the fly the next time it happens.

You will feel, and your associates and managers may feel, that you are going to a lot of trouble when a little experimental tweaking will do the job. Sometimes it will (there is another call on your judgment), but my experience is that scientific method is the cheapest and fastest overall.

I have found that in debugging complex electrical circuits it is more cost-efficient for the engineer to sit at a desk looking at drawings while a technician manipulates the hardware at his or her direction and reports observations and instrument readings by voice. The engineer can concentrate on thinking and analysis.

In Chap. 2 I told you how I analyzed the failure of the sealed-up heart-lung machine by visualizing what was going on inside. It is such visualizing, "thinking like a fish," that you should train yourself to do.

When you experiment, make one change at a time, or you won't know which change caused the effect. That is easy to say. We always hope that the second change will affect something else. Besides, we are under pressure and we are in a hurry. There is a mathematical theory of the design of experiments in which statistical analysis separates the effects of several changes made at the same time or at different times. I have not studied it, but for your particular work it may be worthwhile for you to research it in the library and learn to use it. Reference 1 is an introduction to the subject.

Beware of doing harm when you intend to do good. (Remember "iatrogenic" disorders?)

When you are testing the final product, be sure that it *is* the final product and that later changes will not go to market untested. Be sure that you are not assuming that test results from an earlier model apply to the final model when a relevant change has been made.

The best debugging job I ever did, and the only one which saved a life, was on the Convair heart-lung machine. The story was told in Chap. 2 to illustrate the importance of visualizing to the inventor.

In science, skill in experimentation is different from skill in theoretical work; in engineering, skill in testing and debugging is different from skill in designing in the office.

REFERENCES

1. Rennard, R.: "Statistical Leveraging," *Machine Design,* Aug. 20, 1987.

24

Improving Existing Designs

Much of the design engineering done in the world is in modifying existing designs. It is often much more cost-efficient to improve something which already exists than to start over again from scratch. Many useful kinds of modification are possible. A few of these are listed below.

Scaling

You can scale up or scale down an existing design either to make small improvements in quality or to achieve a major new effect.

Figure 6.1 shows the Decimal Keeper slide rule. It was described in detail in Chap. 2. The significance for the present subject is that a major advance in the product was made merely by changing the scale of the printed pattern on its face.

A second example of the use of scaling and labeling to form a new product is the Cent-R-Liner (Fig. 24.1). In this case a standard half-scale engineering scale was relabeled with zero in the middle and with major divisions running

12 11 10 9 8 7 6 5 4 3 2 1 *0* 1 2 3 4 5 6 7
8 9 10 11 12

If zero is placed on a centerline, then a centered $2^{11}/_{16}$-in width can be marked by marking the $2^{11}/_{16}$-in point on the left and the $2^{11}/_{16}$-in

Figure 24.1   Cent-R-Liner drafting scale.

point on the right without mental arithmetic. Since the second edge was there, it was marked full-sized with zero in the middle so that absolute distance from the centerline could be measured in each direction.

$$6 \quad 5 \quad 4 \quad 3 \quad 2 \quad 1 \quad 0 \quad 1 \quad 2 \quad 3 \quad 4 \quad 5 \quad 6$$

I manufactured and sold both products by mail order for some years.

Reproportioning

A product may be improved by changing its proportions. For example, consider the history of the automobile tire.

Engineering analysis has advanced radically, particularly by using computers with techniques such as finite element analysis and complex number-crunching programs which had been impractical before. It has become possible to proportion and size electrical circuits and components and mechanical structures and parts with much less reliance on judgment and large factors of safety. Old products can now be safely and economically reproportioned by using such analysis.

Adding and Subtracting Features

A product may be improved by adding and subtracting features. We are all familiar with the continual stream of new features in such consumer products as TV sets and automobiles.

A product can be improved by eliminating obsolescent features. Tire pumps are now omitted from automobiles.

(Do you know the engineering distinction between "obsolete" and "obsolescent"? A worn-out product is obsolete; an old-fashioned product is obsolescent.)

Materials

A product can be improved by changing materials. For example, in automobiles aluminum or plastic has replaced steel in many parts. As new materials are developed, each offers an opportunity to improve your product or reduce its cost.

Components

A product may be improved by changing components. As with materials, new components are continually being developed. Many offer opportunities to improve your product.

Manufacturing Methods

A product can be improved by changing the manufacturing methods used in producing it. Machined parts can be changed to die-cast or to powder metallurgy parts. Small changes may make automatic assembly possible. New manufacturing techniques, such as laser machining, are developed occasionally, and some may provide opportunities to improve your product or at least to reduce its cost. It may be necessary to make detail design changes to be able to utilize a new technique.

Performance Ratings

The performance of the product may be increased or decreased. For example, the power of an automobile engine may be decreased to reduce pollution and gasoline consumption. When gasoline is cheaper and catalytic converters are available, the power of the engine may be increased to attract customers.

Combining or Subdividing Parts

Parts and subassemblies may be either combined or divided to reduce manufacturing cost, improve appearance, reduce maintenance cost, increase reliability, or for other reasons. The part change may include a manufacturing-process change. For example, two machined parts may be combined into a single casting. In electronics there is a continuing flood of new integrated circuits which combine previously separate circuits.

Modularizing

The product may be modularized both to reduce manufacturing cost and to permit the generation of a line of products having different features, each provided by its own module. For example, computer systems are extreme cases of modularization, each large computer being a custom combination of standard modules. In the machinery field, my robots were custom combinations of standard modules in which each module produced a single axis of linear or rotary motion. A custom stack of such modules, each made in several sizes, produced a custom machine having the number of degrees of freedom and the load and travel capacities required by its application.

Cosmetics

A product may be modified by changing its appearance and decoration. This is called "industrial design." Often such a change is intended not as a functional improvement but only to make the product conform to current fashion. Automobiles and other consumer products are examples. Clothing design is 99 percent appearance and decoration and 1 percent application of new materials. Athletic shoes are an exception: jogging shoes and ski boots incorporate new functional features.

Reducing Maintenance Needs

A product may be modified in many ways to reduce maintenance. Better bearings, exclusion of dirt, simplified adjustments, easier access, redesign of parts subject to failure—the list is endless.

Reducing Costs

A product can be modified in many ways to reduce costs. A common buzzword for this process is "value engineering." A more cynical expression is "second-guessing." By either term it is a good thing to do.

Adding or Reducing Models and Sizes

A product line can be improved by adding models and sizes. It is economical to use a table of "preferred numbers" to cover a range with a minimum number of sizes. For example, a line of motors may be made in theoretical sizes of 1, 2, 4, 8, 16, 32...hp, rounded to 1, 2, 5, 10, 20, 50...hp. Thus any load can be served without either proliferating sizes or using a grossly oversize machine.

Families of products can be developed with ranges of features and with luxury and utility models. The consumer electronics and automobile industries practice this to extremes.

Export models can be added. This may require a major effort for metrication, conformance with foreign codes, and instructions and nameplates in foreign languages and symbols.

Simplifying

A product can be improved by simplifying it. Chapter 15 is devoted to simplification.

Figure 24.2 is an example of product improvement. The product is a deep-sea camera whose original manual controls were intricate, unreliable, and expensive. I redesigned the internal linkages between the external levers and the lens motions so that they were simpler, cheaper, and more reliable. I proposed and added a cassette system for

Figure 24.2   Deep-sea camera. (*Courtesy of Honeywell Corp.,*
Hydro Products Operations.)

the film which allowed the film to be changed on a sunny deck instead
of the camera having to be taken into a darkroom.

Trade Names

It sounds frivolous, but you can improve the commercial success of a
product by coining trade names for the product and its features. Think
of the products you know which are known by their trade names
rather than by their functional descriptions.

There is no limit to the improvements you can make. The principal
ingredient is really wanting to.

Design Objectives

There is a surprisingly long list of objectives you should bear in mind while developing a design. Some are obvious and some rather less so.

Specifications versus Real Needs

Its seems self-evident that you should meet the explicit specifications spelled out by the customer and contracted for by your company or called for by your marketing department. However, as discussed in Chap. 19, whether these specifications were written in private by your customer or were negotiated with members of your own organization, they recite not what can best serve the customer but what it thinks it can get and thinks it can afford and therefore asks for. When you work on the inventive phase of design, you may reach some different conclusions. Some of the things asked for may turn out to be much more difficult and costly than anticipated. Conversely, some things not asked for may be features which you can invent and offer, which would be useful to the customer, and which it may be able to afford, yet when the customer wrote the specification it didn't say so and didn't ask for them because it did not know.

You may be able to provide features not asked for at substantially less cost than was anticipated by the customer. You may find that some of the things the customer asked for are really rather more costly than was anticipated. The usual position taken by design engineers is to say that the specification is the law and they must conform rigidly to the law. The alternative is to develop an unsolicited proposal, secure the approval of your management, and offer it to the customer as a proposed revision of the specification. If the attitude of all concerned is constructive and is not controlled by a low innovation index, the customer will get more for its money than it expected, your company will be in a more profitable position, and you yourself will be a hero.

Before you stick your neck out, you must form a judgment and a prediction of the response of both your own people and the customer's people. This is the kind of changing of the world which is within your professional capability and in which you have a reasonable chance of success. It is, of course, a keen test of the persuasive power I keep urging you to develop.

A successful example of such an effort was the function generator for the Air Force described in Chap. 19.

If a program is conducted most cost-efficiently, the *specification* will evolve with the evolution of the *design* because the designers will continue to evolve more and better ideas for executing the real purpose of the project. Administratively this is not easy. Even if the specification comes from the marketing department in your own organization, there will be continuous controversy as to what features to add, what features to subtract, what features to change, and what costs to charge. This is an arena in which your own persuasive power has its greatest opportunity.

On the other hand, when you are given an explicit specification by a customer who remains at arm's length for legal reasons (as in competitive bidding) or who is stubborn or when your management insists that no unsolicited proposals be made, then you must conform rigorously to the specification and to the subspecifications in the specification tree.

Abuse Resistance

Your product and its components will be sized to withstand computed stresses, whether electrical, chemical, or mechanical, with some customary factor of safety. This is not enough.

Your product must withstand some degree of abuse, depending on the nature of the environment into which it is sent. Abuse generates transient peak stresses of many kinds rarely explicitly specified except in some military specifications in which shock and vibration tests are specified. Let me list a few examples of real-world abuse which are difficult to quantify.

- A factory machine is run into by a forklift truck.
- An office machine receives a spilled cup of hot coffee.
- A mechanic applies an improper tool such as a hammer instead of a wrench.

Abuse may come from many sources, and a partial list may help you visualize some of the kinds of mistreatment by various persons for which you must design:

- Owner (ignorance, negligence)
- Installer (ignorance, negligence, error)
- Employee (negligence, sabotage)
- Renter (cars, trucks, tools)
- Children
- Hostiles (burglars, vandals)
- Maintainers (negligence, ignorance)
- Public (public telephones, dispensers of gasoline, candy, cigarettes, slot machines, etc.; ignorance, negligence, hostility)

Industrial products for use in factories, transportation facilities, and the like are subject to more abuse than are consumer goods, in part because they are used by people who don't own them. Furthermore, the cost of a breakdown is apt to be greater than the inconvenience of the breakdown of a consumer product (other than a car). Therefore it is customary to design industrial products to be more resistant to abuse and longer-lasting than consumer products. Furthermore, industrial products may be used 8, 16, or 24 hours a day, whereas consumer products typically have very intermittent use except for refrigeration and heating, ventilation, and air conditioning. In the case of consumer products there is an additional problem: there may be more pressure for lower cost and a more artistic appearance than is the case in industrial products. Here the opportunity and the challenge for two-way winners are very great.

Environment Resistance

The environment can be hostile in many ways. Some of these ways exist in service and some in transportation.

- Heat and cold
- Humidity, high and low
- Liquids (water, oil, chemicals, seawater, coffee spills)
- Vapors (smoke, smog, chemicals, salt air, food)
- Organisms (fungi, bacteria, animals)
- Dust and dirt

- Abrasion
- Burns (cigarettes, soldering irons, cookware, stoves)
- Shock and vibration (in service and in shipment)
- Vacuum (space vehicles)
- Radiation [sunlight, electro-magnetic (radio-frequency interference), nuclear]
- Inversion (turning upside down, usually in shipment)
- Voltage surges, low and high voltage
- Electrical noise
- Pressure surges
- Wet and dirty compressed air
- Improper fuel, dirty fuel (I once helped with the victims of a private-plane crash due to engine failure from dirty fuel.)

Environment Protection

Not only must your product not be injured by the environment but it must not injure the environment, particularly the people in it. You must provide guards to protect people from their own errors, clumsiness, and disobedience to instructions. (An injury may be the user's *fault*, but it may also be your *responsibility* for not preventing it.) If the product has a failure, it must not become a fire, injury, or pollution hazard; i.e., it must fail safe. Your product must not pollute the environment with harmful gases, vapors, or liquids during its useful life and should not contain materials which will slowly pollute after its retirement and scrapping. It must not cause slowly developing injuries to its operators. Examples are eyestrain from computer screens and damage to fingers from too frequent operation of difficult push-buttons.

If you can arrange to examine products returned by customers for any reason, you will get a degree of enlightenment which no amount of reading can match.

The list is endless if you visualize the environment in the real world where your product will go. As a result of such histories of abuse, conventional styles of construction have been developed in different industries and in different companies. In the truck business the word "trucky" is used to describe devices which appear to be sufficiently robust that truck drivers and maintainers will feel confident in their prolonged satisfactory performance.

We once built a special robot for major auto company Q. Thinking that I knew the level of negligence and abuse in automobile factories, I made the structure of heavy structural steel about three orders of magnitude stronger and stiffer than there was any functional need for. When the company supervisor visited for a preliminary inspection, he sneered at the construction and said that in his industry it was customary to build machinery *rugged*. Sometimes it pays to build equipment which looks more robust than it really is or needs to be.

Use Preferred Components

It is desirable that you use materials and components which are both standard and preferred in your organization, the customer's organization, or the industry. If you do so, you will please people in each organization and save money for both. If you use uncommon and unconventional materials and components without a need to do so, you will irritate the same people and increase costs.

For aesthetic, functional, and economic reasons, it is desirable to use families of similar components. For example, it is desirable to use the same kind of screwheads or as few variations as possible in a single product.

Maintainability

You should design for maintainability and reliability, whatever those words mean for your product. (See Chap. 22.)

In both consumer and industrial products the maintenance question "Shall we fix it, or shall we replace a disposable module?" is always present. This is not a simple question to answer. It depends on the availability of maintenance skills, the availability of a spare-parts distribution network, and business policies of your company. Business considerations such as the profitability of spare parts and sales of spare modules and a service organization may be a determining consideration. Some companies such as IBM and Xerox generate great confidence in their prospective customers because the customers know that if there is a maintenance problem, skilled help will quickly arrive. Here your marketing department and senior management will have opinions with which you may or may not choose to argue.

Products for the military are subject to design criteria established by the military, although you may have some options within the limits they set. The working environment may be severely hostile (heat, cold, humidity, fungus, shock, vibration, dirt, wet, and, in some cases, negligent and abusive operation). Different ground rules for maintenance are required by different branches of the service. A rifle must be capable of being field-stripped and cleaned by a private soldier under fire in a mud puddle. A Navy shipboard radar probably has a staff of trained and well-equipped technicians in a clean, dry shop.

Be aware of the concept of maintenance echelons, and design for echelon maintenance. The first echelon is in the field and exchanges modules, makes simple adjustments, and otherwise uses relatively little skill, equipment, and time. If first-echelon maintenance does not fix the trouble, the product or its faulty module is sent to second-echelon maintenance, where more skill and equipment are available and individual components may be unsoldered, extracted, etc., and replaced. The top echelon is return to the factory for repair or rebuilding.

Different kinds of service require different maintenance practices and reliability requirements. A newspaper printing press may require extensive daily maintenance by skilled workers, but when it is turned on, it must run perfectly for a few hours, or the cost is disastrous. A space vehicle must operate with no maintenance other than what can be done by remote control. Domestic appliances must work with almost no preventive maintenance, but an occasional random failure is acceptable.

OEM products are made to be sold to manufacturers who incorporate them as components into their own products. Here, ease of installation and testing are of extreme value to your customer and should be designed into your component.

Suit OEM Components to Automation

The increased use of automatic handling, testing, and assembly machines makes it of value to your customer (and therefore to your own organization) to make components easy for automatic feeding and handling in automatic-assembly machines. Among the things you can do to achieve this end is to make the external shape rigid (no dangling wires, tubes, or chains), provide substantial difference in dimensions so that the part feeder does not confuse length with width, and provide clearly usable gripping features.

In the electronics industry, since the late 1950s there has been an increasing practice by component manufacturers of providing their components mounted on reels of tape for feeding into automatic-assembly machines. I designed such a taping machine in 1955. The detailed design of the tape, spacing, etc., must match the handling equipment bought by the OEM from the handling-equipment company. Either industrial standards must be followed where they exist, or all three parties must coordinate their efforts. This practice, which is analogous to feeding a machine gun with a cartridge belt, has been extended down to tiny transistor and integrated-circuit (IC) chips and up to large connector blocks. I have lived with many vibratory part feeders which worked a lot of the time and consider such belt feeding a blessing to all concerned.

Consider proposing a magazine-and-feeder system as an extension of your product line. Several companies whose primary products are fasteners, wire terminals, and adhesives now make and sell such auxiliary products. These in turn make the primary products more desirable.

The most common way of feeding parts to automatic-assembly machines is by vibratory part feeders. Your OEM products must be suitable to withstand the vibration and abrasion, bending, and tangling induced by such feeders if they are to be used. The parts should not tangle and should have distinct features to enable the feeder to orient them. Large parts are sometimes unscrambled by belt feeders, which are slightly less abusive but rather expensive.

Fail-Safe and Fail-Soft

Products should fail safe. Any failure you can imagine happening should result in either a machine stopping or some other situation in which no further damage is done. It is not permissible for a product failure to generate a chain reaction of failures with damage outside the boundaries of the product. Devices such as fuses, circuit breakers, shear pins, slip clutches, a variety of sensor-initiated shutdowns, and other safety precautions should be provided. Machines should stall safely when overloaded.

Another form of protective design is fail-soft. A product with this characteristic develops failures gradually in such a way that the performance of the product deteriorates slowly, thereby giving warning to its operators that maintenance is needed. An elementary example is a knife which gradually gets dull.

Available skill in maintenance and operating personnel is an important consideration in the design of the product. You should design your product to suit both the maintainer and the operator. A family car should be designed differently from an earth-moving machine, in part because one is owner-operated and the other is employee-operated.

Identify Your Company

You must identify your organization as the manufacturer of the product if it is to be sold and not used internally. Therefore you should become acquainted with your company's preferred colors, shapes, styles, markings, and nameplates. If you do not, you may face extensive re-design when the marketing people see your design.

The behavior of OEM customers with respect to their vendor's iden-tification varies widely. I sold a machine-tool-loading robot to ma-chine tool manufacturer T to be incorporated in its display for a na-tional machine tool show. This was a major coup for our small company because there was no other way we could afford to exhibit at this show. When I arrived at the show, I found that all our nameplates had been removed; there was no identification of our company any-where in either the hardware exhibit or the literature; and company T implied in its exhibit posters that the robot was its own product. After a discussion which reverberated through the hall, I was permitted to replace our nameplates on the robot. I was told that it was common in the machine tool trade to chisel off the nameplates of motors so that no name other than that of the machine tool builder would appear on its product.

Most of these war stories tell of troubles of many kinds. I have not included stories of all the products sold to satisfied customers who re-turned to order more, because there are no lessons in them other than "Do good work."

I hope I have stretched your mind somewhat so that you will be on the alert for problems which you can solve before they bounce back from the customer.

Entrepreneuring

There really are many millionaires who made their money by starting businesses based on their inventions. These are the successful entrepreneurs. Among the famous names are Edison, Ford, McCormick, Jobs, and Hewlett and Packard.

The entrepreneur doesn't sell inventions but sells manufactured products instead—a very different ball game. In the end the money comes either from business profits or from selling the operating company as an acquisition to a larger company. I have done this twice on a very small scale, and Jack Rabinow has done it once on a much larger scale.

Now comes the bad part. Most such businesses die young, with a total loss of the money, work, and passion which have been invested. They die because the entrepreneur is not sufficiently competent in the many arts of business other than engineering. (The statement is made that the principal source of failure is insufficient capital. This puts the cart before the horse. The entrepreneur's first responsibility is predicting capital needs and not spending any capital until there is enough. To do otherwise is incompetent.)

On the subject of entrepreneurial incompetence I claim expert status, since I have made most of the mistakes in the book. I managed to reach the end of the tunnel covered with more rose petals than mud, but as Wellington said after Waterloo, "It was a near thing."

Some Advice

Get an M.B.A.

Get an M.B.A. at night while you work as an engineer by day. You will get an introduction to all the business arts I have referred to. Specifically you should learn about:

- Marketing
- Accounting
- Finance
- Marketing
- Business law
- Marketing
- Manufacturing

The repetition of "marketing" is not a typographical error. The better-mousetrap syndrome (which I once shared) is the route to disaster. You will be astounded at the total indifference the world can show to your brainchild.

Have partners

Start with partners who are expert in those business arts in which you are not. Ideally you should have one in accounting, finance, and administration, one in marketing, and one in manufacturing. With yourself in charge of engineering you have a viable core. Ideally these four people have worked together before in some larger company and know that they can cooperate under stress without quarreling.

A trouble with starting on your own without such partners is that you must then recruit managers in these fields. Good people are reluctant to join small, speculative enterprises instead of taking good jobs in big companies. If the enterprise fails, they not only are out of jobs but have bad marks on their records. Many dislike the restrictions of small scale and lack your vision of growth and success to come. In retrospect, I attribute most of my own business troubles to my inability to recruit good managers.

One of the benefits of starting out with a management team is the greater ease of acquiring capital. Investors in start-ups bet on the people at least as much as they bet on the product. A team with good track records is a better bet than a loner with a good track record, to say nothing of a loner with no business record at all.

This is one example of the exponential law of success and failure: each success makes it easier to achieve the next success; each failure makes it more difficult to prevent the next failure.

Become a marketer

The four-partner start-up is an ideal rarely realized. Therefore you must be prepared to multiply your operating responsibilities. I suggest that the most important work you should be prepared for is marketing and sales. (Marketing is strategy, literature, advertising, etc. Sales is going eyeball to eyeball with customers.) It is common for the president of a small company also to be the chief salesperson and to spend much time with customers. That is one of the reasons you need good managers back at the ranch. Furthermore, this is the only reliable way you can know what the market is, how your product is regarded, what your competition is, and what the opportunities are for product improvements and new products. Salespeople are professional optimists and are tempted to lie in the hope that a little more time will turn things around.

For example, my most violent disillusionment was at Typagraph Corp. when I took a consultant's advice and visited my sales representatives around the country. I learned that they had no competence in presenting the product (it was the first computer terminal with hard-copy graphics) and were neglecting it. Nothing was happening except lying.

I learned that innovative complex products must be sold by trained factory employees, not by independent manufacturers' representatives. I had hired a sales manager who had an excellent track record in selling simpler products through sales representatives and was determined to sell this one the same way even after I had given him a direct order to change. That was when I took the plunge into the cold water, the real world of selling my products, and I have never climbed out.

When a manager leaves a big company, there is a transient during replacement, but there are others ready to substitute, at least temporarily. When a manager is lost in a small start-up, the effect can be disastrous. My greatest recruiting triumph was a top man in accounting, finance, and administration. He had been a consultant to me on the side, became a personal friend, and gave up a major job to join my computer terminal company as vice president. Then within a year both his teenage adopted children became hippies, and he was so broken up he became useless.

Raising Capital

The most conspicuous obstacle to starting your own company is the need for capital. The standard sources are yourself, your family, your friends, debt, venture capitalists, and the public.

I started Mobot Corporation on my personal money and a license from Cutler-Hammer Inc. to make the robot system I had developed there. There was the usual desperate struggle, with venture capitalists patting me on the head and saying how interesting this was and that they were going to watch me.

Then, in the same week, independently, *Newsweek, Barron's,* and *Fortune* all printed articles with titles equivalent to "The Robots Are Coming!" Within a week three investment bankers phoned me with offers to take me public; I went with one of them. Blind luck is the biggest component of business success and failure. It also helps to be ready when she knocks on the door.

Your Business Plan

The essential first step in starting a business is writing a business plan. You should do this for your own sake before you take any other steps beyond casual conversations. A business plan is a prediction of the history of the business. It contains the following sections:

1. *Product plan.* What will you sell, how much R&D is required, and how will the product line develop over time?

2. *Marketing plan.* Who constitute the market, why should they buy your product, how many will they buy, how receptive are they to new ideas, what is the competition, how much will it cost to do the selling, how much will customers pay for the product, what will sales be as a function of time? (See Chaps. 13 and 14.)

3. *Manufacturing plan.* How will the product be made, how much tooling investment will be needed, what will it cost to make the product?

4. *Personnel plan.* What jobs will be filled, what individuals do you already have in mind, how many of each job will be filled as a function of time?

5. *Financing plan.* Where will the capital come from, how much, and, if there is to be progressive financing based on achievement, what will the terms be? This part of your plan is a proposal to investors.

6. *Cash-flow spreadsheet.* Cash in and cash out, itemized, by the month for the first year and by the quarter for the next 5 years. A personal computer spreadsheet program will be a big help.

Every potential investor will insist on studying this business plan.

The greatest value of the business plan is that it makes you think out and research what you want to do. You may decide not to do it. If you decide to go ahead, I can guarantee that the written plan will make you change some of your ideas and change them before the changing costs you anything.

I once submitted a very detailed and carefully worked out business plan to a bank to back up a request for a credit line. The wise old banker read, smiled, turned the page, read, smiled, etc., until he finished. I sat with white knuckles. He said, "Very nice job, Mr. Kamm. You understand, I don't believe a word of this because you don't know what will really happen. However, this shows that you have considered all the kinds of things which may happen, so I am confident that you will not be thrown into total panic when the unforeseen occurs. We will give you the line of credit." Was he ever right!

This would be a good time to reread the Nathan Zepell story in Chap. 2 to see an illustration of the cycle of frustrated inventor to entrepreneur to profitable acquisition.

REFERENCES

There are many books written on this subject.
1. Dible, D.: *Up Your Own Organization: A Handbook for Today's Entrepreneur,* Simon & Schuster, New York.
2. Von Siemens, W.: *Inventor and Entrepreneur,* 1st ed., Kelley, New York, 1966.

Designing for Automation

You can reduce the cost and increase the uniformity of your products if you design them so that they can be manufactured with automatic machinery.

Automatic Part Fabrication

Part fabrication by automatic machine requires little difference in design from part fabrication by human-operated machine except for the features required for automatic handling and feeding described below. However, burrs are poison! They jam. A human assembler will deburr, discard bad parts, and re-form bent parts; an assembly machine will not unless you provide extra stations to do so.

Automatic Assembly

Assembly by automatic machine requires substantial modification of a design as made for manual assembly, or the assembly machine will be needlessly expensive and unreliable.

Ten design rules for automatic assembly

1. Make the parts robust enough to withstand handling by machine. Fragile parts can be handled without breakage by careful humans but not by machines.

2. Avoid near symmetry. A machine may have to orient the part,

and it is easier to make a machine discriminate among different orientations if the part is very different in its different orientations.

3. Eliminate shapes which can tangle. Humans can untangle springs, pigtails, and hooks, but machines lack human vision and dexterity. Springs are the worst offenders. Helical springs should have closed-end turns. If possible, they should have nonuniform pitch.

4. Design parts which will come out uniform from their fabricating process. Nonuniform parts jam in feeders and handlers.

5. Visualize the motions the parts go through during assembly, and design them to go together with simple motions. (As an assembly machine designer, I have put parts together with intricate monkey motions and was proud of the achievement, but I'm glad I didn't have to pay for the machine.)

6. Avoid screw, washer, and nut combinations. Washers and nuts have to be held in place by part of the machine while the screw is driven by another part of the machine. Use self-tapping screws or screws with captive washers. Best of all, eliminate fasteners and make the parts snap together or be held together with toy tabs or the equivalent.

7. Minimize part count. For each part there must be a separate increment of assembly machine. A special case occurs when there is more than one of the same part and the machine can assemble all the identical parts at the same station with an indexing portion to move the assembling operation from place to place. Printed-circuit-board assembly is an extreme case of this practice.

8. Where male and female parts engage, provide tapers on one or both parts to guide them together despite small alignment errors in feeding.

9. Consult your manufacturing engineers and part vendors to see if parts can be oriented and loaded into magazines which will then be loaded into the assembly machine. This practice is now widespread in the electronics industry but not spread widely enough. Replacing vibratory feeders with magazine feeders is the best single thing you can do to make assembly machines simpler and more reliable. Furthermore, some of the other rules can be ignored with magazine feeding.

10. Design parts which can be fabricated right at the assembly machine, as they are used. No magazining or vibratory feeding is needed. Small stampings made from strip stock and helical springs which do not need heat treatment or plating are good examples.

Optimum Degree of Automation and Mechanization

Despite its glamour, full automation is not the most cost-efficient solution to every manufacturing problem. It requires the greatest investment of capital, it takes the most time to design and set up, it requires product design modifications to make it possible, and therefore it pays off only for very large quantities. With a computer-controlled machine which can be quickly reprogrammed (including NC machining) individual batches may be small, even one-piece, but total quantity per year must be large to justify the investment in the machine. Furthermore, there are severe limits to what you can do by just reprogramming control computers: reprogramming does not change fixtures or dies, conveyors, feeders, or many other elements necessary to make a product.

Fully automatic machines

There are several types of fully automatic machines:

- Robots, jointed-arm and Cartesian, for machine loading and unloading, for tool handling (welding, deburring, painting) and for some assembling
- Automatic storage and retrieval machines (a class of Cartesian robot)
- Automatic guided vehicles
- Automatic-assembly machines
- Machine tools (NC or template- or cam-controlled)
- Looms, spinning machines, chain makers, headers, and innumerable other specific-product machines

The reprogrammability of robots has been grossly exaggerated by their marketers. Yes, the motions can be changed quickly, but a robot cannot be transferred from one task to a different kind of task without extensive work in installation, change of end effectors, and setup of associated equipment. As with NC machine tools, robots are flexible in *variation within a task* but not in *conversion from task to task*.

Human-controlled powered machines

Such machines include the following:

- Fork trucks. These do the work of automatic guided vehicles and automatic storage and retrieval machines but trade off capital cost for a human driver.
- Load balancers and cranes. These enable a human to manipulate heavy loads without the cost of a robot.
- Machine tools, human-controlled and -loaded or -unloaded.

■ Human-loaded and -unloaded processing machines of many kinds.

Human work with power tools

Power tools in human hands require more skill but cost much less than automatic or semiautomatic machinery. Powered hand tools use electric, pneumatic, or hydraulic power. Some have internal part feeders, such as riveters.

Human work with special hand tools

Surgeons, dentists, and optometrists all have developed specially formed, well-made hand tools to match specific tasks. The variety of hand tools used in manufacturing is small, and rarely does one see a truly clever special tool in use. A special case is automobile maintenance, for which many special tools are made in quantity and sold to garages. I believe that very substantial manufacturing-cost reductions could be made if more attention and ingenuity were brought to bear on hand tools.

28

The Theory of Design

You may encounter the words "theory of design" someday and wonder what you have missed in your education as a design engineer, so I will tell you of my experience with them.

I attended two meetings sponsored by the National Science Foundation to encourage requests for research grants in the theory of design, and I listened to a number of papers and talked with a number of the participants.

All the people I heard and met were academics who were highly trained in mathematical engineering but who had never designed anything that got built and sold in the world of industry. When I was introduced they looked at me in awe and said, "You're a practitioner!" (I had been called many things before, but never a "practitioner.") They meant that I actually designed things.

After cutting through an incredible in-group jargon, I learned that what they really wanted was to do design by computer; not design computation, or drafting, or rule following (as in laying out printed circuit boards or integrated circuits), or data storage and retrieval, but design *thinking*. They really wanted to make science fiction come true by building an artificial brain, just as some robot enthusiasts who have never been in a factory want to build an artificial human worker (with a computer for a brain, of course). I tried to explain some of the things in this book, but they understood me no more than I understood their jargon.

One of the professors, Dr. Billy V. Koen, was trying to interest the others in a system of "heuristics," or pragmatic principles of design, but it wasn't a computer program, so they ignored him.

Nothing is all bad. Those visits gave rise to my course, and the course gave rise to this book.

The Human Interface

Importance

Every product must be dealt with by a human, if only to turn it on. It is surprising in how many ways a product which is easy and clear to its designer can be difficult or confusing to the person who receives it. Many misfortunes, small and large, have resulted from bad design in this respect, and much pleasure and satisfaction have resulted from good design. The first major study of what is now called "human engineering," "human factors," or, to be fancy, "ergonomics," was made in the design of the airplane cockpit, where the pilot has little time in which to respond to many inputs with many outputs and there is a terrible price for error.

Many books and articles have been written on the subject, but they are rarely included in undergraduate programs. This chapter is a brief survey of the subject.

Simplest Rule

Imagine yourself in every possible relationship between a human and your product, and ask yourself how you could miss or misunderstand what is going on, how you could do the wrong thing at any time because of ignorance or confusion, and how you could make it easier to see, hear, and do what must be done.

Fourteen Specific Rules and Principles

1. You should conform to conventions and not be "creatively" different unless you have very good reason. The Three Mile Island disas-

ter in 1979 was due in part to the use of red lights to indicate normal conditions and green lights to indicate trouble! The poorly trained operators reacted to the usual (reverse) meaning of the colors and overrode perfectly well-operating automatic safety controls to generate the catastrophe.

2. Your product should not injure operators or bystanders either catastrophically or gradually. Provide guards, safety switches, interlocks, warning lights, color patterns, and whatever else you can think of to prevent an ignorant, careless, or tired person from being hurt mechanically, electrically, or chemically. Prevent gradual injury from noise, bad lighting, or frequent tolerable stress. (There is something called "tunnel carpal syndrome" which comes from operating pushbuttons or switches too frequently under certain conditions. At the time of writing, there is a major product liability lawsuit by postal workers who have this syndrome from operating the input switches which set up the zip code for each letter on the automatic sorting machine I have described elsewhere. I'm glad I had nothing to do with those switches!)

3. A word about safeties and interlocks on production machines: operators try to disable or bypass them to make their jobs faster and more convenient. In effect there is a contest between design engineer and operator to save or lose fingers and hands. One of my customers who made plastic-molding machines told me of an operator who bypassed four concentric safeties by reaching over the top of the machine to clear a problem: one hand gone. (The customer wanted to put my robots on his machines to prevent this sort of thing, but his chief engineer and sales manager persuaded him that his customers would not pay for them.)

Many years ago I made an experimental die and tested it on a manual press. It worked fine, so we gave it to the factory. The factory added a rudimentary guard and put it on a power press without showing it to me: one fingertip. I got cold shivers for years while wondering how I could have been responsible, but there was no way I could have been.

4. The product should be physically easy to operate. Forces and torques should be within easy effort of the weakest people who will operate it. Reach required should be within convenient distances for small people.

5. We depend on touch as well as sight to identify controls and position, especially in the dark or when our eyes are fixed to something else. Among my favorite bad designs are the brake release and the hood release on many cars, each of which is a T handle to be pulled and which are located close to each other. The electronic and aircraft

industries have large families of knobs of different sizes, shapes, and textures to prevent errors.

There has been a rash of automobile accidents in which a car has surged forward when the driver wanted it to stop. The Audi was the most publicized example. No mechanical defect was found. The cause was attributed to the closeness in position and feel of the brake pedal and the accelerator pedal so that the driver stepped harder and harder on the accelerator, thinking to press harder on the brake. It's true. My wife almost killed a woman in exactly that way with our Audi while I was a passenger. We missed the woman, who didn't even see us, but we almost died of fright. (The recall fix was brilliant: Audi designed and added a mechanical interlock which prevented shifting out of park or neutral unless there was pressure on the real foot brake. And this was on a retrofit on recalls!)

6. Color and color patterns (e.g., black and yellow stripes) are useful in identifying the meaning of signals, identifying control knobs, buttons, and levers, and warning of unsafe regions and objects. I don't know how to handle the problem of color blindness.

A large set of symbols has been standardized for international use to convey messages without words.

7. Illumination has been going on for so long and in so many places that it is remarkable how badly it is sometimes done. There is an Illuminating Engineering Society which will be glad to send you literature. You should avoid lamps which shine directly into your eyes, glass or polished surfaces so oriented that they reflect light directly into your eyes, and very nonuniform brightness of displays. (Your eyes accommodate to the brightest object in view. One of my pet peeves is candlelight in restaurants. Your eyes are attracted to the bright flame, the pupils contract to suit it, and you cannot see anything else.)

8. An operator should be comfortable enough so that there is no fatigue which would slow responses but not so comfortable that sleep is likely. Similarly, the operator should not be so busy that fatigue is likely but not so idle that important signals or observations are missed.

9. The number of information displays is important: too much information is fatiguing and masks important information, while too little fails to give the operator what is needed. There is an opportunity, in complicated products, to use blinking, brightness change, color change, warning lights and sounds, and other means to attract the operator's attention to urgent information.

10. Whenever an operator moves a control, there should be a response so that the operator knows he or she has succeeded in the con-

trol action. Joysticks on airplanes with power-driven control surfaces are provided with actuators to generate artificial resistance so that the pilot feels as if he or she is flying an airplane with directly coupled controls. Pushbuttons should move in response to pressure, or there should be a light or sound or other feedback so that the operator knows a contact has been made.

11. Inputs and indicators for different portions of a system should be grouped together.

12. The choice of digital or analog displays is largely a matter of fashion. (Elsewhere I tell of my friend the physicist who promotes computer graphics because of the ease of understanding analog displays but who insists on digital displays for his automobile dashboard.) Where resolution of three significant figures or more is required, only digital displays are easily read.

13. Products should fail safe; that is, failure should result in a stopped rather than a runaway condition or should not otherwise result in human hazard. Furthermore, it is desirable that products should fail soft; that is, they should deteriorate in a noticeable way before they fail altogether so that the operator is given notice of imminent total failure before it happens.

14. Not only should products resist harming humans but they should be resistant to harm by humans. Chapter 22 discusses abuse resistance. One form of such resistance is "foolproofing," i.e., resistance to erroneous operation which is not malicious.

There are professional consultants who will be glad to critique and improve your human engineering, but some knowledge on your own part during initial design will save much time and money doing redesign because of their advice.

Human engineering yields better cooperation between product and people.

REFERENCES

1. McCormick, E., and M. Sanders: *Human Factors in Engineering and Design,* 6th ed., McGraw-Hill, New York, 1987.
2. Woodson, W.: *Human Factors Design Handbook: Information and Guidelines for the Design of Systems, Facilities, Equipment, and Products for Human Use,* 1st ed., McGraw-Hill, New York, 1981.
3. ———and D. Conover: *Human Engineering Guide for Equipment Designers,* 2d ed., University of California Press, Berkeley, 1964.

Chapter 30

Approximations

Real structures and circuits are usually too complicated to compute exactly. Modern computers with programs such as finite element analysis have made enormous breakthroughs in computing the performance of physical systems. However, even they require some simplifying assumptions, and they are time-consuming to apply to the early design of a new product. Therefore it is useful to develop skill in approximation and simplifying assumptions. Then, as the design converges to its final form, either pencil and paper or computer analysis, or both, can be used for more exact performance and sizing calculations.

Insight

The most basic requirement for approximating with minimum error is an insight into the behavior of the product and its components. This is a glittering generality which is easy for me to say but impossible for me to explain. Nevertheless, it is true, and you should consciously try to imagine the spread of stress and strain, the flow of currents and generation of voltages, and the motion of material in your mental image of the product. Just as you make and improve a mathematical model inside a computer, you should make and continually improve an imagination model inside your head.

Seven Principles of Approximation

1. One of the techniques of approximation is the substitution of lumped parameters for distributed parameters. For example, in a

structure loaded with spread loads you can substitute concentrated loads at what you judge to be appropriate positions. In electrical devices it is common to make such approximations; for example, the distributed leakage inductance and capacitance of a transformer winding is almost perfectly represented by a lumped inductance in series and a lumped capacitance in parallel with a leakage-free winding. (In the case of high-frequency transients such as lightning strokes on a power transformer this approximation breaks down.)

2. You can approximate the device itself by omitting details and by replacing complex portions with approximately equivalent simple portions.

3. Your approximation does not even have to conform to the laws of nature. For example, Prof. Herbert C. Roters of the Stevens Institute of Technology developed a technique of computing the reluctance of a magnetic field by approximating the shape of the field with a set of geometric shapes each having an average length and average width and adding the parallel reluctances of the shapes.[1] Figure 30.1 illustrates the approximation for the field between two pole pieces. For the approximation to be "true" there would be points of infinite flux density, yet the calculations match experimental measurements with surprising accuracy.

4. The roughest approximation is used in testing the practicality of the earliest conception of the product. Then, as the design is developed and refined, the degree of approximation is reduced.

5. There are advanced CAD programs which will do almost exact analysis of structures and devices as they are sketched into the terminal. If you have and can use such a computer and software, then you need do little approximation.

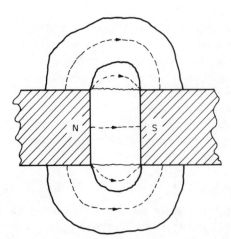

Figure 30.1 Magnetic-field approximation.

6. A word on approximating for mental or back-of-the-envelope arithmetic: when considering a new idea, you can get a very useful insight with one-significant-figure arithmetic. Round off everything to one significant figure. Assume 10 in to the foot, 3 for pi, 4000 seconds to the hour, 400 days or 10 months to the year, and so forth. In dealing with very large or very small quantities use the exponential form, e.g., 4×10^4 instead of 43,290. Your results will be good to only one significant figure, but the calculation will be very fast and the one-figure result will lead you to decisions which will result in more useful designs worthy of more accurate calculations.

7. In Chap. 2 I spoke of the undesirability of exact definitions and categories in the early-thinking phase of your work. The use of such undefined words is a qualitative approximation.

This chapter is not an appeal for inaccuracy; it is an approach to efficiency so that you can reach maximum useful accuracy with the least time and work.

REFERENCES

1. Roters, Herbert C., *Electromagnetic Devices,* Wiley, New York, 1941, 1944.

Chapter 7

Minimum Constraint Design
(MCD)

What It Does

This chapter deals with design principles for achieving, simply and economically, mechanisms with zero looseness, zero binding, and zero stress due to assembly. The same principles produce static structures which can be assembled with zero looseness and zero interference, due to tolerances, during assembly. The principles are old, but I have met few mechanical engineers who are acquainted with them. I have used these principles for years and trained my staffs to use them with great benefit to our companies and clients.

Reference 1, which was my source of education in the subject, uses the expression "kinematic design" for these principles. This expression suggests the design of moving linkages, which it does not mean here, so I use the expression "minimum constraint design," which I think is more descriptive. (Unfortunately Ref. 1 is now out of print, but you may be able to find it in a library or a used-book shop.)

Basic Theory

An unconstrained rigid body has six degrees of freedom, which I will call X, Y, and Z linear freedoms and roll, pitch, and yaw rotary freedoms. If it is constrained at one point, it retains five degrees of freedom; if it is constrained at two points, it retains four degrees of freedom; and so on, until if it is constrained at six points, it has zero freedom. If an attempt is made to constrain it at more than six points, the body will be forced to deform; i.e., there will be binding or interference or internal stress from the deformation needed to conform to the overconstraint. This sounds pretty abstract, so let's take an example.

An Example

In Fig. 31.1, rigid body A is mounted on rigid body B on three legs, C, D, and E. Each leg has a hemispherical tip. Assume that all parts and shapes are made with loose manufacturing tolerances.

Leg C seats in conical hole 7. Since the leg end and the hole both have finite shape errors, they will touch at three points, 1, 2, and 3. If you imagine holding A in your hand, you will see that A now is constrained in X, Y, and Z but that it remains free to move in roll, pitch, and yaw. Three points of constraint have removed three degrees of freedom.

Now seat leg E in V-shaped groove 8. It will touch at points 4 and 5. Body A now has only one free rotation left, about the axis joining the

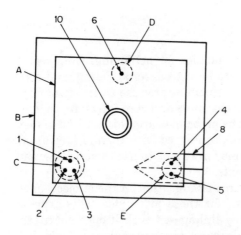

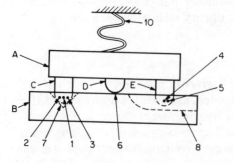

Figure   31.1 Six-point constraint.

hemispheres at the ends of legs C and E. There are five points of constraint, five degrees of freedom constrained, and one degree of freedom unconstrained.

Now rotate A about that C–E axis until leg D touches the flat surface of B. There are six points of constraint and zero degrees of freedom.

Add spring 10 to hold the two parts together. The spring provides a *force* but does not add a *constraint* because the spring end is flexible in all six axes.

I have built working devices with exactly this structure, and the effect is weird. The parts go together and come apart with no effort whatever; yet when they are together, they have zero looseness and feel as if they were welded.

Compare this with a conventional doweled assembly.

Heavy Loads

Lightly loaded structures can be made in exactly this way, but heavy loads will Brinell the points of contact. This may not be bad: the cold-worked contact areas will still be nearly perfect in effect and yet have finite contact areas. Even lightly loaded "point" contacts elastically form contact areas as in a ball bearing or an electrical relay contact.

For heavily loaded structures you compromise with the theoretical purity of your point contacts in a variety of ways. The cone seat 7 can be a spherical seat of slightly larger radius than the C leg end so that the contact stresses under high load are tolerable. The tip of leg E can be wedge-shaped and free to rotate about the E axis to give a comparable effect. The tip of leg D can sit on an intermediate block with a spherical hollow on top and a flat below with the same effect.

You can devise many other forms of replacing the theoretical six points with six areas and still have the same elegance of zero looseness, zero binding, and loose tolerances.

If you are designing a mechanism with moving parts, you do exactly the same thing but you use five or fewer constraint points, depending on the number of degrees of freedom you want for each part.

Using Commercial Components

A very valuable exception to pure MCD consists in making the interfaces between bodies commercial components made with close-tolerance multiple internal constraints. Among these are:

- Rolling bearings
- Spherical rod ends

- Precision ways with matching linear self-aligning bearings
- Tooling balls and separate balls
- Spherical washer pairs
- Tooling pressure pads with internal spherical joints
- Universal joints
- Flexible couplings

The only field of design which I know of that uses minimum constraint practice regularly is machine-shop fixture design. The fixture-design (or tooling-design) component companies make many products which can be used in a variety of structures and mechanisms to provide small area constraint points.

I had an amusing consulting experience in using this art. I visited an electronics company which had mounted an instrument servomechanism on the same board as its control circuitry. The rotating parts were very tight (i.e., binding), and the company's technicians could not make them turn freely. Cables were held down with many cable clamps, and, to match, shafts were held down with many bearings. I told them to do nothing but remove certain bearings. Instant fix. Client awe.

In my robot company we made machines with 150 ft of horizontal travel and 20 ft of vertical travel by using only a few simple machine tools and no machining of large areas. We shipped without fully assembling them in our own plant because our shop was too short. They went together in the field without trouble and with zero looseness and zero binding.

I have found that the use of these principles is a skill which grows with practice. I strongly recommend that you do practice it. You will achieve both functional and economy success.

REFERENCES

1. Whitehead, T. N.: *The Design and Use of Instruments and Accurate Mechanisms*, 1st ed., Dover, New York, 1954. Out of print.

Design for Packaging and Shipping

Necessity

The packaging and the shipping plan are parts of the product design and are as much the responsibility of the designer as are the working parts of the product. You should design the product to be adapted to shipping, and you should consider packaging and shipping to be unwritten parts of the specification. (Some customers, including the military, make them very much a written part of the specification.)

Appearance

The appearance of a consumer product package may be an important sales tool. The product itself may be displayed through a transparent window in the package. Such packaging is a job for an expert. Unless you have become one, you should call on your own company's packaging department, if it has one, or on one or more packaging vendors as soon as your product's size has been established. If you use an industrial designer (artistic), he or she should be authorized to work on both product and package as an artistic whole.

Protection

The first requirement of packaging is that it protect the product from damage during unsympathetic handling from your factory to the end user. Such handling includes dropping from truck to ground, usually on a corner, rain, cold and heat (including standing in the hot sun), shock and vibration in trucks and freight cars, low pressure on airplanes, salt spray on ships, dust, impact from other packages, electromagnetic radiation including x-ray examination, and careless unpacking by the end user or its receiving department.

Among the techniques used to protect the product are partial disassembly and separate packaging or subpackaging of the parts, cushioning with corrugated cardboard or cut or molded foam cushions, plastic film bags, silica gel packets enclosed to absorb moisture, coating corrodible or polished surfaces with greases or peelable plastic films, corrosion-protecting bag materials, and blocking or clamping of flexible structures. You should consider what the customer should do if the product may have to be repacked for further shipment or return.

Naked Shipment

A valuable technique to avoid the cost of packaging protection is naked shipment, calling on the upbringing of all but the crudest brutes to make them care for the product. During World War II there was much breakage of radar display tubes. Finally they were shipped naked, suspended in flimsy openwork orange crates. Everyone has been brought up not to drop glass.

For large products there has been developed the air-ride van, similar to a standard furniture moving van. The product is loaded at the factory, unpackaged, with straps and cushioning quilts, by the factory's own people working with the driver. The driver has been selected and trained by the moving van company to handle sensitive and expensive products. The same driver picks up the product, drives it all the way without transshipment, and delivers it to the end user. Most of the large moving companies provide this service. The cost per ton-mile is high, but the cost for crating and uncrating is almost zero and damage risk is almost zero. I used to ship very large robots in this way with complete success. I told my customers that it was the most expensive but cheapest form of transportation I could find.

Sterile Packaging

For certain medical products and, of course, for certain food products, the package must be capable of being sterilized by either heat or gas. Experiments have been made with x-ray and nuclear radiation for sterilizing after packaging.

For drug products the package must be tamperproof or tamper-evident against some of the less lovable members of the community and against children.

Free Work from the Customer

Some packaging systems transfer final assembly to the customer, thereby simplifying the package and saving assembly labor in the factory.

Documentation

A most important part of customer documentation is unpacking instructions. Remember that customers can be most obtuse about understanding a written instruction that they have never seen before and must obey accurately. Many customers follow the rule: "When all else fails, read the instructions." Therefore you should visualize a customer tinkering with assembly. Design the product so that the customer will do the least harm.

Packaging Vendors

There is a large industry of custom package companies. They will design and provide cushions, cartons, boxes, pallets, and crates to suit your product. Treat them as you would any special-component vendor.

There is another industry of custom *packaging* companies. They will pack, crate, and mark your product to meet a variety of specifications including military, customs, and those of other countries.

An old anecdote about Henry Ford in the Model T days is his specifying the dimensions of the wooden crates in which vendors sent him parts. He used the crate wood for floorboards.

33

Nonengineering Design

You have talent, training, and experience as a design engineer. You use these professionally to design airplanes, amplifiers, or chemical plants. You can also use them, to your advantage or at least to your entertainment, to design many other things which are not considered to be "engineering." Here are some examples.

Contract Terms

You make contracts to buy or lease apartments, houses, and cars. You make bank loans. The terms, which are presented to you as "standard" to discourage you from negotiating them, are only as standard as the other person's bargaining position permits. For example, I once learned, the hard way, to add a buy-out clause to apartment leases so that I could leave at any time (usually because of a job change) by paying one month's extra rent. Car dealers are always interested in making deals. Negotiate redesigns of your contracts! If you become an entrepreneur, the possibilities to design favorable contracts with vendors, customers, financiers, and employees are endless.

Political deals. You will have political problems in your job (Chap. 3). You can sometimes design an informal deal in which both parties benefit. (In every deal you make in life, be sure that both parties benefit, or there will be trouble no matter how ironclad the contract may be.)

Home Architecture

Sooner or later you will renovate a house or an apartment. You would be surprised at what you can come up with in innovative and convenient features if you bring your design capabilities to bear.

Artistic Industrial Design

No, you are not an artist, but you have seen many good-looking products and many bad-looking ones. You may have better taste than you realize. *Try* your hand at the artistic design of your product. If you do not use an industrial designer, it will look better than if you did not, and if you do use an industrial designer, you will give him or her a head start and reduce the amount of functional redesign he or she will ask you to do to accommodate the artistic work.

Marketing Plans

Yes, that is the marketing department's job, but if you get to do proposal work, a product proposal with a marketing plan you have designed will sound more like a winner to your management than one without.

Business Plans

These plans are primarily for the entrepreneurs among you, but that includes in-house entrepreneurs who propose a new business area to their present company. (See Chap. 26 for details.)

Proposals

The core of a proposal is the design of something new. It will usually be the design of a new product, but it can include a development plan, a test-marketing plan, a financing plan, and anything else you can design which will tempt the customer to buy.

Language

Whenever you produce a sentence, you have designed a combination of words. This is traditionally the most hateful field of design to engineers, but neglecting it costs us dearly. Please reread Chap. 4, "Persuasion: The Golden Art."

Administrative Matter

This includes procedures, forms, and paperwork. When and if you enter management, you will find plenty of opportunity to redesign such stuff, preferably by making it simpler. It's enormously satisfying to do so. One of my most satisfying achievements was a no-work expense report.

Engineering Documentation

Documentation comprises wiring, parts lists, notes, standards, and the like. When you computerize, if you have not yet done so, there is a tremendous opportunity to redesign your system. I once replaced all our wiring drawings with computerized wiring lists. There was a double saving: engineering costs were less and wiring costs were less because the lists were easier to read and handle and there were fewer errors.

I have done design in each of these categories with great satisfaction and considerable profit. (See Chap. 2, "Inventing," for more fields of design and invention.) I am sure that in your own life there are opportunities not listed here. Go to it!

And now, goodbye. I wish you a successful career.

Index

ABOUT THE AUTHOR

Larry Kamm's formal education includes a B.S.E.E. from Columbia in 1941 and an M.E.E. from Brooklyn Polytechnic Institute in 1946. He has Professional Engineer licenses from New York, Maryland, and California and is a registered patent agent and a certified manufacturing engineer (SME). He belongs to Sigma Xi and the Institute for Electrical and Electronics Engineers (IEEE) and, when he worked in those fields, to the American Rocket Society [now the American Institute of Aeronautics and Astronautics (AIAA)], the Society of Manufacturing Engineers (SME), and the Robot Institute. He holds 37 patents, issued or pending, and has published or presented 26 papers and much trade literature.

Kamm has invented and designed mechanical and electro-mechanical devices of great diversity. These include robots, numerical controls, computer peripherals, space vehicles and components, simulators, a mail sorter memory, a heart-lung machine, automatic test-and-sort and other manufacturing equipment, transducers, switchgear, and engine components.

He has worked as an employed engineer in both small and large companies, as an entrepreneur (Numerical Control Corp., Devonics, Inc., Typagraph Corp., Mobot Corp.), as a teacher of design theory and practice, and as a consulting inventor and designer, which is his present activity. He lives in San Diego, California.